Michael Masuch   László Pólos (Eds.)

Knowledge Representation and Reasoning Under Uncertainty

Logic at Work

Springer-Verlag

Lecture Notes in Artificial Intelligence   808

Subseries of Lecture Notes in Computer Science
Edited by J. G. Carbonell and J. Siekmann

Lecture Notes in Computer Science

Edited by G. Goos and J. Hartmanis

808

Lecture Notes in Artificial Intelligence

Subseries of Lecture Notes in Computer Science
Edited by J. G. Carbonell and J. Siekmann

Lecture Notes in Computer Science
Edited by G. Goos and J. Hartmanis

Michael Masuch   László Pólos (Eds.)

Knowledge Representation and Reasoning Under Uncertainty

Logic at Work

Springer-Verlag
Berlin Heidelberg New York
London Paris Tokyo
Hong Kong Barcelona
Budapest

Series Editors

Jaime G. Carbonell
School of Computer Science, Carnegie Mellon University
Schenley Park, Pittsburgh, PA 15213-3890, USA

Jörg Siekmann
University of Saarland
German Research Center for Artificial Intelligence (DFKI)
Stuhlsatzenhausweg 3, D-66123 Saarbrücken, Germany

Editors

Michael Masuch
László Pólos
Center for Computer Science in Organization and Management
University of Amsterdam
Oude Turfmarkt 151, 1012 GC Amsterdam, The Netherlands

CR Subject Classification (1991): I.2.3-4, F.4.1

ISBN 3-540-58095-6 Springer-Verlag Berlin Heidelberg New York
ISBN 0-387-58095-6 Springer-Verlag New York Berlin Heidelberg

CIP data applied for

Typesetting: Camera ready by author
SPIN: 10131227   45/3140-543210 - Printed on acid-free paper

Preface

"Twenty years ago, logic was mainly applied to mathematical and philosophical problems. Nowadays, the term applied logic has a far wider meaning as numerous applications of logical methods in computer science, formal linguistics and other fields testify. Such applications are by no means restricted to the use of known logical techniques: at its best applied logic involves a back-and- forth dialogue between logical theory and the problem domain. Ultimately, these applications may change the face of logic itself. A variety of non-standard logics (for example modal, temporal and intuitionistic logics, lambda calculi) have gained increased importance, and new systems (such as update semantics, dynamic logic, and various non-monotonic logics) have emerged in response to the new questions."

This was the key paragraph in the invitation for the international conference Logic at Work, organized jointly by the Center for Computer Science in Organization and Management (CCSOM) and the Institute of Logic, Language, and Computation (ILLC) and held at the University of Amsterdam, Dec. 17-19th, 1992. The conference attracted 86 submissions from all over the world, out of which the program committee (Patrick Blackburn, Jean-Jules Meyer, Frank Veltman, plus the editors of this volume) selected 30 papers with the help of anonymous referees. In addition, there were 8 invited speakers (e.g., Peter Gärdenfors, Yoav Shoham, and Petr Hájek). Out of these contributions, we selected 13 papers related to two core issues of formal AI for publication in a volume on knowledge representation and reasoning under uncertainty.

The Dutch National Science Foundation provided the funding for this workshop through a PIONIER-grant. Anja Krans maintained the day-to-day communication with the authors, and processed the text together with Breanndán Ó Nualláin. Henk Helmantel kept the computers up and running. Thanks are due, last but not least, to all participants of the conference, and in particular to the head of the local arrangements committee, the organizer extraordinary Michel Vrinten.

April 1994   Michael Masuch and László Pólos

Table of Contents

The Role of Expectations in Reasoning

Peter Gärdenfors

Lund University, Cognitive Science,
Kungshuset, Lundagard,
S-223 50 Lund, Sweden
E-mail: Peter.Gardenfors@fil.lu.se

Abstract. Logical theory traditionally assumes that (1) logical inference is a relation between sentences (or propositions), not between thoughts (or anything cognitive), and (2) the validity of an argument depends solely on the logical structure of the sentences involved and is independent of their meaning. In practical reasoning, however, these assumptions are not valid. In this paper I want to show that by taking expectations into account, one can achieve a much better understanding of how logic is put to work by humans. In particular, one obtains a very simple analysis of nonmonotonic reasoning. I will also discuss the cognitive origins of expectations.

Then a man said: Speak to us of Expectations. He then said: If a man does not see or hear the waters of the Jordan, then he should not taste the pomegranate or ply his wares in an open market. If a man would not labour in the salt and rock quarries, then he should not accept of the Earth that which he refuses to give of himself. Such a man would expect a pear of a peach tree. Such a man would expect a stone to lay an egg. Such a man would expect Sears to assemble a lawn mower.

Kehlog Albran, *The Profit*

1 Ubiquitous Expectations

We all have expectations. For example, you presumably expect a speaker at a scientific meeting to be standing properly on his feet while talking. You may also expect that he is not red in his face, although this expectation may be much weaker than the previous one.

Sometimes, however, our expectations are violated. If a speaker is standing on his hands while giving his presentation, you would, I venture, be surprised. But given this violation of your previous expectations, you would now expect him to be red in his face, even if you don't observe it. This example shows that we are able to reason and draw conclusions even when our primary expectations are gainsaid.

As a matter of fact, expectations are ubiquitous, although they are not often made explicit. You expect there to be a floor when you enter a room; you expect a door handle not to break when you press it; you expect your morning newspaper to arrive on time; and you don't expect Sears to assemble a lawn mower. The main thesis of this article is that expectations play a crucial role in everyday reasoning. In brief, we use them as supplementary premises when we make inferences. In particular, I will argue that much of *nonmonotonic logic* can be reduced to classical logic with the aid of an analysis of the expectations that are hidden in the arguments. I will also discuss the cognitive origins of expectations.

In classical logic, the role of expectations is eschewed. Why is this so? For an answer, we should first reconsider some of the philosophical and methodological assumptions that have governed the traditional approach to logic since Frege.

Classical assumption 1: Logical inference is a relation between sentences (or propositions), not thoughts (or anything else related to cognition).

According to the traditional view, logical arguments are described as a relation between a set Γ of premises and a conclusion C, in symbols $\Gamma \models C$. The premises and conclusions are expressed as sentences in some language, preferably a formal language.
What kind of relation is "$\models$"? Traditionally this is answered as follows:

Classical assumption 2: The validity of an argument is only dependent on the logical structure of the sentences in Γ and C and independent of their meaning, their truth or the context.

The inference relation $\models$ can be specified in two major ways: (1) proof-theoretically, by specifying the axioms and derivations rules that generate $\models$;[1] or (2) semantically, by formulating the truth conditions for the logical relation to hold. In either case, assumption 2 is valid.

However, expectations, as opposed to axioms or standard premises, are *defeasible*. If an expectation is in conflict with a premise, it yields. Consequently, expectations don't fit with the classical assumptions.

Hence expectations are suppressed in classical logic, but in practical reasoning they are not. Everyday arguments are full of hidden premises that need to be made explicit in order to make the argument logically valid. In each separate case, it may be possible to add the hidden assumptions to make the derivation comply with the classical assumptions. But nobody does this, because expectations are normally shared among speakers, and unless they are countervened, they serve as a common background for arguments.

If we want to describe a logical inference relation that better conforms with everyday reasoning, we need a notion of inference that is not constrained by the classical assumptions. But if these assumptions are abandoned, we must give an answer to the question: *what can be used as premises in an argument?* For classical logic, the answer is: only the set Γ. For practical reasoning the answer may be different. As I shall argue in Section 3, the answer to this question may help us understand much of what has been called nonmonotonic logic.

---

[1] Some of the axioms may be 'non-logical', which normally means that they are taken from a 'theory'.

2 Luria's Camels

Before turning to an analysis of the role of expectations in nonmonotonic reasoning, let me present an extreme case (extreme from the classical point of view that is) of handling premises. Consider the following dialogue taken from (Luria 1976: 112):

Subject: Nazir-Said, age twenty-seven, peasant from village of Shak-himardan, illiterate.

> The following syllogism is presented: **There are no camels in Germany. The city of B. is in Germany. Are there any camels there or not?**
> Subject repeats syllogism exactly.
> **So, are there camels in Germany?**[2]
> "I don't know, I've never seen German villages."
> *Refusal to infer.*
> The syllogism is repeated.
> "Probably there are camels there."
> **Repeat what I said.**
> "There are no camels in Germany, are there any camels in B. or not? So probably there are. If it's a large city, there should be camels there."
> *Syllogism breaks down, inference drawn apart from its conditions.*
> **But what do my words suggest?**
> "Probably there are. Since there are large cities, there should be camels."
> *Again a conclusion apart from the syllogism.*
> **But if there aren't any in all of Germany?**
> "If it's a large city, there will be Kazakhs or Kirghiz there."
> **But I'm saying that there are no camels in Germany, and this city is in Germany.**
> "If this village is in a large city, there is probably no room for camels."
> *Inference made apart from syllogism.*

On the basis of this and a number of similar interviews, Luria draws the conclusion that illiterate people are far inferior to literates when it comes to logical reasoning.[3]

I don't agree with his diagnosis. I believe that what is at stake here is exactly the issue of what is allowed as premises in an argument.

In contrast to Luria, my hypothesis is that the peasants that he is interviewing do not allow themselves to use anything as a premise for an argument unless it is part of their *personal experience*. So in the dialogue above, Nazir-Said *ignores* the information provided in the syllogism, since he has no direct

---

[2] Luria's text probably contains an error here. The question should have been "Are there any camels in B.?"

[3] Also cf. (Luria 1979)

knowledge of the matter ("I don't know, I've never seen German villages)".

In my opinion, the point becomes very clear in a dialogue like the following (Luria 1976: 108-109):

> The following syllogism is presented: **In the Far North, where there is snow, all bears are white. Novaya Zemlya is in the Far North and there is always snow there. What colors are the bears there?**
> "There are different sorts of bears."
> *Failure to infer from syllogism.*
> The syllogism is repeated.
> "I don't know: I have seen a black bear, I've never seen any others... Each locality has its own animals: if it's white, they will be white; if it's yellow, they will be yellow."
> *Appeals only to personal, graphic experience.*
> **But what kind of bears are there in Novaya Zemlya?**
> "We always speak only of what we see; we don't talk about what we haven't seen."
> *The same.*
> **But what do my words imply?** The syllogism is repeated.
> "Well, it's like this: our tsar isn't like yours, and yours isn't like ours. Your words can be answered only by someone who was there, and if a person wasn't there he can't say anything on the basis of your words."
> *The same.*
> **But on the basis of my words - in the North, where there is always snow, the bears are white, can you gather what kind of bears there are in Novaya Zemlya?**
> "If a man was sixty or eighty and had seen a white bear and had told about it, he could be believed, but I've never seen one and hence I can't say. That's my last word. Those who saw can tell, and those who didn't see can't say anything!" (At this point a young Uzbek volunteered, "From your words it means that bears there are white.")
> **Well, which of you is right?**
> "What the cock knows how to do, he does. What I know, I say, and nothing beyond that!"

The upshot is that Luria's evidence does not indicate that there is anything wrong with the logical abilities of the illiterate peasants. The main difference is that they don't take a statement provided by a stranger as something that can be used in reasoning; *only personal experience is allowed* (there seem to be strong moral feelings about this among the Uzbeks).

The illiterate Uzbeks violate the second classical assumption of traditional logic since the validity of an argument is extremely dependent on the personal experience of the person who presents the conclusion, to the extent that the explicit premises in Γ may be totally ignored. The notion of logical validity of illiterate Uzbeks is quite different from the classical one. This does, however, not mean that Luria's illiterate Uzbeks are illogical - it only means that they are reasoning by other rules.

Luria's interviews show that when people learn to read, they will learn to play a different logical game. My understanding, which does not completely fit with Luria's, is that literate individuals learn to see a text as an abstract entity, independent of a particular speaker and his practical experience or his motives for uttering the words spoken. They see a text as a symbolic structure. It is only in relation to such a symbolic structure that the classical assumptions make sense. For somebody who only hears spoken words, uttered by a particular person, it is much more difficult (and of little practical importance) to view the words as symbolic structures. In other words, literacy teaches us to separate abstract arguments from their practical context.

I have presented my analysis of Luria's interviews as an extreme case of deciding which premises may be used in a logical argument - the case when even the premises explicitly stated in Γ may be disregarded. Be this as it may, the point of the example is that one must be careful in specifying what counts as a premise for an argument when practical reasoning is concerned

3 Nonmonotonic Reasoning Based on Expectations

In some recent articles, David Makinson and I have argued that the areas of nonmonotonic logic and belief revision are very closely related (see: Makinson and Gärdenfors 1990, Gärdenfors 1990, Gärdenfors 1991a, and Gärdenfors and Makinson, to appear). In particular, we show in (Gärdenfors 1991a), and (Gärdenfors and Makinson, to appear) how various forms of nonmonotonic inferences can be given a unified treatment in terms of how *expectations* are used in reasoning. This section begins with a summary of our analysis, but also discusses some limitations of the assumptions employed there.

3.1 Motivation

When we try to infer whether C follows from Γ, the information we use for the inference does not only contain the premises in Γ, but also background information about what we *expect* in the given situation. For instance, if we know that someone is a Spanish woman, we anticipate her to be dark and temperamental. Such expectations can be expressed in different ways: by default assumptions, by statements about what is normal, or typical, etc. These expectations are not premises that have to be accepted, but they are *defeasible* in the sense that if the premises Γ are in conflict with some of the expectations, we don't use them when determining whether C follows from Γ.

I want to show that expectations are used basically in the same way as explicit premises in logical arguments; the difference is that the expectations are, in general, more defeasible than the premises.[4] Consequently, the expectations

---

[4] But cf. the examples from Luria above.

used in nonmonotonic inferences need no special notation, but they can be expressed in the same language as regular premises and conclusions. This is one side of the unified treatment of nonmonotonic reasoning. For simplicity I shall work with a standard propositional language L which will be assumed to be closed under applications of the *boolean connectives* ¬ (negation), ∧ (conjunction), ∨ (disjunction), and → (implication). I will use α, β, γ, etc. as variables over sentences in L. I will assume that the underlying logic includes *classical propositional logic* and that it is *compact*. Classical logical consequence will be denoted by $\models$, and the set of classical consequences of a set Γ will be denoted $Cn(\Gamma)$.

In this section and the following, all the different expectations will be formulated in L. In contrast to many other theories of nonmonotonic reasoning, there are thus no default rules or other additions to the basic language, such as modal operators, that will be used to express the defeasible forms of information. Another, non-propositional, way of handling expressions will be presented in Section 6.

The key idea behind nonmonotonic reasoning can be put informally as follows [5]:

α *nonmonotonically entails* β *iff* β *follows logically from* α *together with "as many as possible" of the set of our expectations as are compatible with* α.

In order to makes this more precise, we must, of course, specify what is meant by "as many as possible".[6] But before turning to technicalities, let me illustrate the gist of the analysis by a couple of examples. "α nonmonotonically entails β" will, as usual, be denoted $\alpha \mathrel{|\!\sim} \beta$.

As a first example, let the language L contain the following predicates:

Sx: x is a speaker at a conference
Hx: x is standing on his hands
Rx: x is red in the face

Assume that the set of expectations contains $Sb \rightarrow \neg Rb$ and $Sb \wedge Hb \rightarrow Rb$, for all individuals b. Assuming that the set of expectations is closed under logical consequences it also contains $Sb \rightarrow \neg Hb$ and, of course, the logical truth $Sb \wedge Hb \rightarrow Sb$. If we now learn that b is a speaker at a conference, that is Sb, this piece of information is consistent with the expectations and thus we can conclude that $Sb \mathrel{|\!\sim} \neg Rb$ according to the recipe above.

On the other hand, if we learn both that b is a speaker and, surprisingly enough, is standing on his hands, that is $Sb \wedge Hb$, then this information is

---

[5] I will restrict the analysis to the case where there are only finitely many premises which can be conjoined to a single α. However, as shown by (Freund et al. 1990), there is a canonical way of extending any such finitary relation to cover infinite sets of premises.

[6] This idea is related to the idea of 'minimal change' within the theory of belief revision (see: Gärdenfors 1988: 66-68).

inconsistent with the set of expectations and so we cannot use all expectations when determining which inferences can be drawn from $Sb \wedge Hb$. The most natural expedient is to give up the expectation $Sb \rightarrow \neg Rb$ and the consequence $Sb \rightarrow \neg Hb$. The contracted set of expectations which contains $Sb \wedge Hb \rightarrow Rb$ and its logical consequences, in a sense contains "as many as possible" of the sentences in the set of expectations that are compatible with $Sb \wedge Hb$. So, by the general rule above, we have $Sb \wedge Hb \mathrel{\vdash\!\!\!\sim} Rb$. This shows that $\mathrel{\vdash\!\!\!\sim}$ is indeed a nonmonotonic inference operation.

3.2 Expectation orderings

Expectations act as hidden assumptions. However, when evaluating their role in arguments, it is important to note that our expectations about the world do not all have the same strength. For example, we consider some rules to be almost universally valid, so that an exception to the rule would be extremely unexpected; other rules are better described as rules of thumb that we use for want of more precise information. An exception to the latter type of rule is not unexpected to the same degree as in the former case. In brief, our expectations are all defeasible, but they exhibit varying *degrees of defeasibility*.

In order to make these ideas more precise, I shall assume that there is an ordering of the sentences in L. '$\alpha \leq \beta$' should be interpreted as 'β is at least as expected as α' or 'α is at least as surprising as β'. '$\alpha < \beta$' will be written as an abbreviation for '*not* $\beta \leq \alpha$' and '$\alpha \approx \beta$' is an abbreviation for '$\alpha \leq \beta$ and $\beta \leq \alpha$'.

According to the key idea of this section, $\alpha \mathrel{\vdash\!\!\!\sim} \beta$ means that β follows from α together with all the propositions that are 'sufficiently well' expected in the light of α. How well is 'sufficiently well'? A natural idea is to require that the added sentences be strictly more expected than $\neg\alpha$ in the ordering. This is the motivation for the following definition.

Definition 1. $\mathrel{\vdash\!\!\!\sim}$ is an *expectation* inference relation iff there is an ordering $\leq$ satisfying $(E1) - (E3)$ such that the following condition holds:

$(C \mathrel{\vdash\!\!\!\sim})$ $\alpha \mathrel{\vdash\!\!\!\sim} \gamma$ iff $\gamma \in Cn(\{\alpha\} \cup \{\beta : \neg\alpha < \beta\})$

To a large extent, the formal properties of the nonmonotonic inference relation defined in this way depends on the properties that are assumed to hold for the ordering $<$. (Gärdenfors and Makinson, to appear) assume that it satisfies the following postulates:

$(E1)$ If $\alpha \leq \beta$ and $\beta \leq \gamma$, then $\alpha \leq \gamma$ (*Transitivity*)
$(E2)$ If $\alpha \models \beta$, then $\alpha \leq \beta$ (*Dominance*)
$(E3)$ For any α and β, $\alpha \leq \alpha \wedge \beta$ or $\beta \leq \alpha \wedge \beta$ (*Conjunctiveness*)

The first postulate on the expectation ordering is very natural for an ordering relation. The second postulate says that a logically stronger sentence is always less expected. From this it follows that the relation $\leq$ is reflexive. The third constraint is crucial for the representation results proved in (Gärdenfors and Makinson, to appear), but presumably the one that is most open to query. It concerns the relation between the degree of expectation of a conjunction $\alpha \wedge \beta$ and the corresponding degrees of α and β.

Note that the three conditions imply *connectivity*: either $\alpha \leq \beta$ or $\beta \leq \alpha$. For by $(E3)$ and $(E2)$, either $\alpha \leq \alpha \wedge \beta \leq \beta$, or $\beta \leq \alpha \wedge \beta \leq \alpha$ and we conclude connectivity by $(E1)$. From $(E2)$ it also follows that $\alpha \wedge \beta \leq \alpha$ and $\alpha \wedge \beta \leq \beta$, so $(E3)$ entails that $\alpha \wedge \beta \approx \alpha$ or $\alpha \wedge \beta \approx \beta$. This means that we cannot interpret the degrees of expectation directly in terms of their *probabilities*, since $(E3)$ is violated by any probability measure. The word 'expectation' as it is used in this paper should hence not be confused with the notion of 'expected utility' in decision theory. 'Expected utility' has to do with expectations of the *values* of various outcomes, while the notion of expectation studied here concerns *beliefs* about the world. In my opinion, this use of 'expectation' comes much closer to the everyday use.

Recalling that by the three conditions on expectation orderings we have $\neg\alpha \leq \beta_i$ for all $i \leq n$ iff $\neg\alpha \leq \beta_1 \wedge \ldots \wedge \beta_n$, it is immediate, using the compactness of Cn, that $(C \hspace{1pt}\vdash\hspace{-6pt}\sim)$ is equivalent to:

$(C \hspace{1pt}\vdash\hspace{-6pt}\sim')$ $\alpha \hspace{1pt}\vdash\hspace{-6pt}\sim \gamma$ iff either $\alpha \models \gamma$ or there is a $\beta \in L$ with $\alpha \wedge \beta \models \gamma$ and $\neg\alpha < \beta$

This condition may be surprising. It says that γ follows from α if there are some expectations that are consistent with α which together with α *classically* entails γ. In other words: *Nonmonotonic logic is nothing but classical logic if relevant expectations are added as explicit premises!* I believe that this observation can remove a lot of the mystery surrounding nonmonotonic inferences. If the analysis presented here is correct, a lot of the paraphernalia of nonmonotonic logic will not be required anymore. Among other things, one needs no new notation for defaults[7], no special inference rules for nonmonotonic logics, and no particular model theory.

3.3   Weaker assumptions about expectations

(Gärdenfors and Makinson, to appear) prove that expectation inference relations, which are based on expectation orderings fulfilling $(E1) - (E3)$, satisfy a number of postulates for nonmonotonic inferences. One of the strongest postulates is the following:

Rational Monotony: If $\alpha \hspace{1pt}\not\vdash\hspace{-7pt}\sim \neg\beta$ and $\alpha \hspace{1pt}\vdash\hspace{-6pt}\sim \gamma$, then $\alpha \wedge \beta \hspace{1pt}\vdash\hspace{-6pt}\sim \gamma$.

This postulate cannot be proved without assuming that the strong condition $(E3)$ holds for the expectation ordering. Conversely, in the completeness proof for expectation inference relations based on orderings fulfilling $(E1) - (E3)$, the proof that $(E3)$ is fulfilled makes essential use of Rational Monotony.

---

[7] Defaults will be analyzed in terms of expectations in the following section.

Like ($E3$), Rational Monotony is a strong postulate, the validity of which is sometimes challenged.[8] For example, (Ginsberg 1986) presents a counterexample to a corresponding principle for conditionals which can also be used against Rational Monotony. His example involves the following statements:

α: Verdi is not French.
β: Bizet is French.
γ: Satie is French.
δ: Bizet and Verdi and compatriots.
ε: Bizet and Satie are compatriots.

In the example, α, β, and γ are seen as background facts, that is, sentences that have a strong degree of expectation (and we may assume that they have roughly the same degree of expectation).

Now, let us first assume δ as a premise and consider the nonmonotonic consequences of this assumption. Since δ is inconsistent with the conjunction of α and β, at least one of these expectations must be given up. However, it is not certain that β is given up, and consequently one cannot conclude $\neg\varepsilon$ from δ, so that we have $\delta \not\hspace{-2pt}\sim \neg\varepsilon$.

Furthermore, since δ and γ are independent statements, the addition of δ does not affect the validity of γ. Hence it is reasonable to suppose that $\delta \hspace{2pt}\sim \gamma$.

Next, let us start from $\delta \wedge \varepsilon$ as a premise, i.e., assume that all three composers are compatriots. In this inferential situation, at least one of the expectations α, β, or γ must be rejected. However, since they are assumed to be of equal strength, it might be that γ is rejected. Consequently, it does not hold that $\delta \wedge \varepsilon \hspace{2pt}\sim \gamma$.

Summing up, we have $\delta \hspace{2pt}\sim \gamma$ and $\delta \not\hspace{-2pt}\sim \neg\varepsilon$, but not $\delta \wedge \varepsilon \hspace{2pt}\sim \gamma$, which gives us a counterexample to Rational Monotony, and, indirectly, a counterexample to ($E3$).

The intuitive validity of the Ginsberg's counterexample depends on the fact that δ and γ are *independent* statements. The notion of independence is difficult to formalize,[9] but in the present context it can at least be stated that independence entails absence of expectations in the sense that if δ and γ are independent, then $\delta \hspace{2pt}\sim \gamma$ if and only if $\hspace{2pt}\sim \gamma$ and $\gamma \hspace{2pt}\sim \delta$ if and only if $\hspace{2pt}\sim \delta$.

As noted above, it follows from ($E3$) that the expectation ordering is total. However, the notion of independence requires that some sentences not be comparable with regards to their degree of expectation. Since the validity of ($E3$) is tightly connected to the validity of Rational Monotony, this is then the cause of the violation of Rational Monotony in Ginsberg's example. For these reasons, it interesting to study weaker versions of ($E3$) that do not entail that the expectation ordering be total, but, say, only that it be a partial ordering. In particular, (Rott 1992) investigates the following two principles:

($E3$ $\uparrow$) If $\alpha < \beta$ and $\alpha < \gamma$, then $\alpha < \beta \wedge \gamma$

($E3$ $\downarrow$) If $\alpha \wedge \beta < \beta$, then $\alpha < \beta$

---

[8] For a defence, see (Lehmann and Magidor 1992).
[9] However, see (Gärdenfors 1978) for a general analysis and (Gärdenfors 1991b) for an application of this analysis to belief revision processes.

In connection with belief revision procedures, he is able to show some representation theorems involving these principles. Neither of his theorems utilizes the equivalence of Rational Monotony (i.e., the postulate K^*8 from (Gärdenfors, 1988)).[10] It remains an open question whether his results can be transferred to the context of nonmonotonic inferences.

4 Defaults as Expectations

One of the main motivations for studying nonmonotonic reasoning is that this kind of theory is necessary if we want to understand reasoning by *default assumptions*. In this section I want to show that an ordering of expectations contains enough information to express, in a very simple way, what we require with respect to default information.[11] The principal idea is that a default statement of the type 'F's are normally G's' can be expressed by saying that 'if something is an F, then it is less expected that it is non-G than that it is G'. This formulation is immediately representable in an expectation ordering by assuming that the relation $Fb \rightarrow \neg Gb < Fb \rightarrow Gb$ holds for all individuals b.

To illustrate the general idea of expressing defaults of the form 'F's are normally G's' as an expectation relation $Fb \rightarrow \neg Gb < Fb \rightarrow Gb$ for all individuals b, assume that all we know about b is that Fb. We want to decide the nonmonotonic consequences of this fact. It can be determined, via $(C \mathrel{|\!\sim})$, that $Fb \mathrel{|\!\sim} Gb$. It can also be determined that $Fb \mathrel{|\!\not\sim} \neg Gb$. Further information about b, for example that Hb, will mean that we no longer need to check whether $Fb \rightarrow \neg Gb < Fb \rightarrow Gb$, but rather whether $Fb \wedge Hb \rightarrow \neg Gb < Fb \wedge Hb \rightarrow Gb$, which may give a different answer. This is exactly how we want a default rule to operate.

To give an analysis of a familiar example, the so-called Nixon diamond, suppose that L contains the following predicates:

Rx: x is a republican
Qx: x is a quaker
Px: x is pacifist

Assume that we have the default rules "republicans are normally not pacifists" and "quakers are normally pacifists." According to the rule given above, we express these defaults by a number of ordering relations of the form $Rb \rightarrow Pb < Rb \rightarrow \neg Pb$ and $Qb \rightarrow \neg Pb < Qb \rightarrow Pb$, respectively, for various individuals b.

From this we conclude, as above, that if all we know about McCarthy is that he is a republican, then we expect him to be a non-pacifist (and we don't expect him to be a quaker); and if all we know about Fox is that he is a quaker, then we expect him to be a pacifist (and don't expect him to be a republican).

---

[10] For a survey of some of the results concerning other principles for expectation orderings, also cf. (Gärdenfors and Rott, to appear).

[11] Cf. (Morreau 1992) for a related analysis of defaults.

Now, suppose that, contrary to our expectations, Nixon is a both a quaker and a republican, that is $Qa \wedge Ra$. What can be concluded concerning his pacifism?

If we know that $Qa \wedge Ra$, and we want to decide whether $\neg Pa$ or Pa follows nonmonotonically, then this can be determined via $(C \hspace{0.5mm}\vdash\hspace{-1mm}\sim)$ by looking for the strictly greater of $Qa \wedge Ra \rightarrow \neg Pa$ and $Qa \wedge Ra \rightarrow Pa$ in the expectation ordering. Three cases are possible:

(1) $Qa \wedge Ra \rightarrow \neg Pa < Qa \wedge Ra \rightarrow Pa$. In this case, we conclude that $Qa \wedge Ra \hspace{0.5mm}\vdash\hspace{-1mm}\sim Pa$.

(2) $Qa \wedge Ra \rightarrow Pa < Qa \wedge Ra \rightarrow \neg Pa$. For similar reasons, we conclude that $Qa \wedge Ra \hspace{0.5mm}\vdash\hspace{-1mm}\sim \neg Pa$.

(3) $Qa \wedge Ra \rightarrow Pa \approx Qa \wedge Ra \rightarrow \neg Pa$. In this case (or in the case when they are incomparable, if the ordering is not supposed to be total), then neither $Qa \wedge Ra \hspace{0.5mm}\vdash\hspace{-1mm}\sim \neg Pa$, nor $Qa \wedge Ra \hspace{0.5mm}\vdash\hspace{-1mm}\sim Pa$ will hold.

None of these three possibilities is ruled out by the two ordered pairs $Ra \rightarrow Pa < Ra \rightarrow \neg Pa$ and $Qa \rightarrow \neg Pa < Qa \rightarrow Pa$. The reason is that it follows from $(E2)$ that $Ra \rightarrow \neg Pa \leq Qa \wedge Ra \rightarrow \neg Pa$ and that $Qa \rightarrow Pa \leq Qa \wedge Ra \rightarrow Pa$. Consequently, the maximum of $Qa \wedge Ra \rightarrow \neg Pa$ and $Qa \wedge Ra \rightarrow Pa$ will be at least as high as each of $Ra \rightarrow \neg Pa$ and $Qa \rightarrow Pa$ in the expectation ordering. But on the other hand, the two comparisons do not suffice to determine which, if any, of $Qa \wedge Ra \rightarrow \neg Pa$ and $Qa \wedge Ra \rightarrow Pa$ is the greater. So, the information available does not permit us to conclude anything concerning $\neg Pa$ or Pa.

To sum up, the nonmonotonic consequences one can draw from the premise that $Qa \wedge Ra$ depends on which is chosen to be the maximal element of $Qa \wedge Ra \rightarrow \neg Pa$ and $Qa \wedge Ra \rightarrow Pa$ in the expectation ordering. The default relations $Ra \rightarrow Pa < Ra \rightarrow \neg Pa$ and $Qa \rightarrow \neg Pa < Qa \rightarrow Pa$ are not sufficient to determine this choice.

5 "But"

We have expectations, but we are sometimes surprised. The choice of the word "but" instead of "and" in the previous sentence indicates that the information contained in the second half of the sentence *violates our expectations*. This, I want to argue, is the core meaning of "but".

In introductory courses in logic, one often uses the formalism of propositional logic to analyse the conjunctions of natural language. It is shown how words like "and", "or", "not", "if...then", "unless", "even if", etc. can be expressed in formulas. But "but" is seldomly given a proper analysis. At best it is said that it has the same logical meaning as "and". For any user of language, it should be obvious that this is false. Among linguists, it is commonplace that "but" expresses a violation of expectations.[12] However, they leave it at that, since they have no way of representing and analysing expectations.

Using the tools of the previous two sections, I want to propose the following:

(*C*But) *A sentence of the form "α but β" is acceptable in a context C if and only if α and β are both acceptable in C, and in C it holds that* $\alpha \mathrel{|\!\!\sim} \neg\beta$.

I don't propose any truth conditions for "but", simply because I don't think there are any. As will be clear soon, the use of but is very context sensitive, so I believe the proper analysis of "but" should be in terms of the conditions under which a sentence is *accepted* in a given context. The context also determines what the current expectations are.

Let me apply the analysis to some examples:

(1) She is rich and ugly.
(2) She is rich but ugly.

The difference between the content of (1) and (2) is that in (2) the speaker presupposes that rich women are normally not ugly, while such an expectation is not indicated in (1). The use of "but" *signals* an expectation. However, which expectation is signalled is not determined by the sentence alone, but depends on the whole context.

The role of the context can sometimes be quite subtle. Compare the following two sentences from Robin Lakoff (1971: 133):

(3) John hates ice-cream but so do I.
(4) John hates ice-cream but I like it.

On a standard reading of (3), it carries the expectation that people normally like ice-cream. The "but" indicates no contrast between me and John, but a contrast between me and people in general. However, in a natural context where (4) is uttered, there is no expectation concerning how normal people like ice-cream. On the other hand, mentioning John's dislike for ice-cream creates an expectation, albeit a weak one, that I too should dislike it, and it is this "inductively generated" expectation that is denied by the "but". The same argument applies to another of Lakoff's examples:

(5) John is tall but Bill is short.

She calls examples like (4) and (5) instances of "semantic oppositions" and argues that it is one of the functions of "but" to express such oppositions (Lakoff 1971: 133). However, I find it more natural to view sentences (4) and (5) as a special kind of denial of expectations, which is the other meaning of "but" that Lakoff identifies and which is expressed more formally in my analysis above.

In this section, I have outlined an analysis of the meaning of "but" based on expectations. In addition to this application, I believe that a large part of

---

[12] For example (Lakoff 1971). She distinguishes (p. 133) between two main meanings of "but": (1) semantic opposition as in "John is tall but Bill is short"; and (2) denial of expectation as in "John is tall but he's no good at basketball". However, as I will argue later, the semantic opposition meaning is a special case of violated expectation.

the discussion within linguistics and philosophy concerning *presuppositions* of sentences can be given a more unified treatment in terms of the expectations of the speaker. However, such an analysis will not be attempted here.

6 How are Expectations to be Represented?

Expectations have, so far, been treated as primitive notions. But where do they come from? In this final section I will discuss the origins of expectations and alternative ways of modelling them.

In Section 3, expectations were modelled by expectation orderings which are orderings of propositions. An important epistemological question for the analysis presented in that section is how this ordering is determined. In Gärdenfors and Makinson (to appear) it is shown that it is possible to define an expectation ordering by using a nonmonotonic inference operation by the following equation:

$$(C \leq) \qquad \alpha \leq \beta \text{ iff either } \alpha \wedge \beta \in Cn(\vee) \text{ or } \neg(\alpha \wedge \beta) \not\vdash \alpha.$$

The case when $\alpha \wedge \beta \in Cn(\vee)$ is just the limiting case when $\alpha \wedge \beta$ is logically valid. The main case when $\neg(\alpha \wedge \beta) \not\vdash \alpha$ means basically that if $\alpha \wedge \beta$ is expected and we assume that $\neg(\alpha \wedge \beta)$, then α is no longer expected, which is the criterion for α being less expected than β.

In (Gärdenfors and Makinson, to appear), we prove in Theorem 3.3 that if $\vdash$ is any inference relation that satisfies the full set of postulates, including Rational Monotony, then the ordering $\leq$ defined by $(C \leq)$ is indeed an expectation ordering over L that satisfies $(E1) - (E3)$.

However, this results does not give a satisfactory solution to the problem of the origin of an expectation ordering – it is like putting the cart in front of the horse. The proposed definition is worthless from a methodological point of view since the nonmonotonic inferences are what is to be explained with the aid of expectations.

A more constructive answer is to view expectations as emerging from *learning processes*. In our roles as cognitive agents, we do not simply observe the world around us, but we also *generalize* in several ways, by discovering patterns and correlations, by forming concepts, etc. Generalizations breed expectations. Expectations are in this way accumulated by *inductive* methods rather than by deductive reasoning. In an evolutionary perspective, expectations can be regarded as a way of summarizing previous experiences in a cognitively economical way.

The analysis presented in Section 3 represents expectations by an ordering of *propositions*. However, if expectations are created by inductive methods, propositional representations of expectations need not be the most appropriate form. In (Gärdenfors, to appear a), I argue that there are three levels of inductive reasoning: *The symbolic, the conceptual, and the subconceptual.* On the symbolic level, inductive inferences are represented by propositions, while on the conceptual level observations and inductive processes are represented by *conceptual spaces* consisting of a number of quality dimensions. On the subconceptual level, finally, observations are described in terms of the perceptual receptors of the mechanism (human, animal, or artifical), which perform the inductive generalizations. In contrast to traditional philosophy of science and AI approaches, I argue that the most important aspects of inductive processes are to be found on the conceptual and subconceptual levels. Consequently, the origins of expectations should be sought on these levels too.

A currently popular method of modelling processes on the subconceptual level is by using *neural networks*. When a neural network is trained, the weights of the connections between the neurons are changed according to some learning rule. The set of weights of a network obtained after the training period is thus an implicit representation of the "expectations" of the network.

In this context, it can be noted that (Balkenius and Gärdenfors 1991) show that by introducing an appropriate schema concept and exploiting the higher-level features of a "resonance function" in a neural network, it is possible to define a form of nonmonotonic inference relation. It is also established that this inference relation satisfies some of the most fundamental postulates for nonmonotonic logics. The upshot is that a large class of neural networks can be seen as performing nonmonotonic inferences based on the expectations of the network. The construction presented in (Balkenius and Gärdenfors 1991) is an example of how symbolic features can emerge from the subsymbolic level of a neural network.[13]

However, neural networks constitute only *one* way of modelling expectations. Apart from their role in logical reasoning, it seems to me that the notion of expectation is central to many cognitive processes. Hence, it is of great interest for cognitive science to investigate different models of expectations. With the exception of "expected utility", the concept does not seem to be much studied within cognitive psychology.[14]

One further exception is Dubois and Prade's (1991) work on the connections between expectation orderings and *possibility logic*, which points in a different direction. In conclusion, I would like to recommend that the notion of expectation be studied from a variety of approaches within cognitive science. There are numerous potential applications of such studies.

Acknowledgments

Research for this article has been supported by the Swedish Council for Research in the Humanities and Social Sciences. Earlier versions of this paper have been presented at the conference on *Logic at Work*, Amsterdam, December 17-19, 1992, and the *European Conference on Analytical Philosophy*, Aix-en-Provence, April 23-26, 1993. I wish to thank the participants at these meetings, in particular Dov Gabbay and Fiora Pirri, for helpful comments.

---

[13] For further discussion of of how logic emerges from the dynamics of information, see (Gärdenfors, to appear b).

[14] And expected utility has to do with expectations of *values*, not expectations about *knowledge* as studied in this paper.

References

Balkenius, C., and P. Gärdenfors: "Non-monotonic inferences in neural networks," in: J.A. Allen, R. Fikes, and E. Sandewall (eds), *Principles of Knowledge Representation and Reasoning: Proceedings of the Second International Conference KR'91*, (San Mateo, CA: Morgan Kaufmann), (1991) 32-39.

Dubois, D., and H. Prade: "Epistemic entrenchment and possibility logic," in: *Artificial Intelligence* 50, (1991) 223-239.

Freund, M., D. Lehmann and D. Makinson: "Canonical extensions to the infinite case of finitary nonmonotonic inference relations," in: G. Brewka and H. Freitag (eds), *Arbeitspapiere der GMD no. 443: Proceedings of the Workshop on Nonmonotonic Reasoning*, (1990) 133-138.

Gärdenfors, P.: "On the logic of relevance," in: *Synthese* 37, (1978) 351-367.

Gärdenfors, P.: *Knowledge in Flux: Modeling the Dynamics of Epistemic States*. Cambridge, MA: The MIT Press, Bradford Books, 1988.

Gärdenfors, P.: "Belief revision and non-monotonic logic: Two sides of the same coin?," in: L. Carlucci Aiello (ed.), *ECAI 90: Proceedings of the 9th European Conference on Artificial Intelligence*. London: Pitman Publishing (1990) 768-773.

Gärdenfors, P.: "Nonmonotonic inferences based on expectations: A preliminary report," in: J. A. Allen, R. Fikes, and E. Sandewall (eds), *Principles of Knowledge Representation and Reasoning: Proceedings of the Second International Conference*. San Mateo, CA: Morgan Kaufmann (1991a) 585- 590.

Gärdenfors, P.: "Belief revision and relevance," in: *PSA 1990*, Volume 2, (1991b) 349-365.

Gärdenfors, P.: "Three levels of inductive inference," in: *Proceedings of the 9th International Congress of Logic, Methodology, and Philosophy of Science*. Amsterdam: North-Holland, to appear a.

Gärdenfors, P.: "How logic emerges from the dynamics of information," in: J. van Eijck and A. Visser (eds), *Logic and the Flow of Information*, to appear b.

Gärdenfors, P., and D. Makinson: "Nonmonotonic inference based on expectation," in: *Artificial Intelligence*, to appear.

Gärdenfors, P., and H. Rott: "Belief revision," in: D. Gabbay (ed.), *Handbook of Logic in AI and Logic Programming, Volume IV: Epistemic and Temporal Reasoning*, Chapter 3.2. Oxford: Oxford University Press, to appear.

Ginsberg, M. L.: "Counterfactuals," in: *Artificial Intelligence* 30, (1986) 35-79.

Lakoff, R. , "If's, and's, and but's about conjunction," in: R. Cole and J. L. Morgan (eds), *Studies in Linguistic Semantics*, New York: Academic Press, Vol. 3, (1971) 114-149.

Lehmann, D., and M. Magidor: "What does a conditional knowledge base entail?", in: *Artificial Intelligence* 55, (1992) 1-60.

Luria, A.: *Cognitive Development: Its Cultural and Social Foundations*. Cambridge, MA: Harvard University Press, 1976.

Luria, A.: *The Making of Mind - A Personal Approach to Soviet Psychology*. Cambridge, MA: Harvard University Press, 1979.

Makinson, D., and P. Gärdenfors: "Relations between the logic of theory change and nonmonotonic logic," in: G. Brewka and H. Freitag (eds), *Arbeitspapiere der GMD no. 443: Proceedings of the Workshop on Nonmonotonic Reasoning*. (1990) 7-27. Also in: A. Fuhrmann and M. Morreau (eds), *The Logic of Theory Change*. Berlin: Springer-Verlag, Lecture Notes in Artificial Intelligence no. 465, (1991)185-205.

Morreau, M.: *Conditionals in Philosophy and Artificial Intelligence*, Ph. D. Dissertation, Universiteit van Amsterdam, 1992.
Rott, H.: "Preferential belief change using generalized epistemic entrenchment," *Journal of Logic, Language and Information* 1, (1992) 45-78.

On Logics of Approximate Reasoning

Petr Hájek

Institute of Computer Science
Academy of Sciences of the Czech Republic
182 07 Prague 8, Czech Republic
E-mail:hajek@uivt.cas.cz

Abstract. A logical analysis of reasoning under conditions of uncertainty and vagueness is presented, using many valued and modal logics and their generalizations. The emphasis is on the distinction between degrees of belief and degrees of truth. The paper is mainly a survey of the relevant results.

1 Introduction

Approximate reasoning is one of the most fascinating branches of AI, and it has generated an extensive literature. Its main notions are *uncertainty* and *vagueness*. They are themselves vague and used in various connotations. We shall associate the term uncertainty with *degree of belief* regarding a proposition (which itself is crisp and may be true or false); on the other hand, vagueness (fuzziness) is associated *degree of thruth* of a proposition (which may be fuzzy, i.e., admits non-extremal degrees of truth). We shall be careful not to mix uncertainty with unprovability of a proposition (given a theory) or with inconsistency (of a theory).

We shall argue that logical systems corresponding to fuziness are many-valued logics, whereas systems corresponding to uncertainty (i.e., degree of belief) are related to various generalizations of modal logics. Furthermore we shall show that presence of both fuzziness and uncertainty gives rise to many-valued modal logics. The point of distinction is whether or not the logical system in question should be *truth-functional* (extensional), i.e., whether the value of a formula is determined by the values of its components.

The study of approximate reasoning has various important aspects: logical aspects (syntax, semantics, deduction, completeness), algebraic aspects (various algebraic structures involved, their algebraic properties), and - last but not least - probabilistic aspects. Clearly, probability theory is a very prominent theory of uncertainty and the notion of uncertainty should be explicitly related to this theory; on the other hand, this should not mean that alternative groundings for the notion of uncertainty are impossible or forbidden.

Our aim is to isolate and survey logical calculi emerging in approximate reasoning, classify them, and state their main properties. The paper is organized as follows:

In Section 2 we introduce the basic distinction between truth-functional and non-truth-functional calculi accomodating vagueness and uncertainty. Section 3 is a digression or rule-based systems. Section 4 deals with many-valued calculi and fuzzy logics, Section 5 surveys basic facts on modal logics, Section 6 deals with calculi of beliefs, Section 7 discusses the combination of vagueness and uncertainty (many-valued modal logics), and Section 8 presents some conclusions and open problems. Thanks are due to the organizers of the meeting "Logic at work" for their kind invitation.

2 Basic Distinction

The distinction between *graded truth values* of fuzzy propositions and *degree of belief* of crisp propositions has been stressed by various authors (see e.g. (Dubois-Prade 1988, 1990, 1991), (Godo and Lopez de Mantaras, to appear)). Both truth degrees of fuzzy proposition and belief degrees of crisp propositions are coded by reals from the unit interval $[0,1]$ (in most cases; we shall not discuss exceptions); but they are handled differently. Notorious examples (like "John is young") show that fuzzy propositions are often used in the way of qualitative speaking about quantities; mostly one deals with fuzzy propositions extensionally, i.e., discusses *how* to define the truth degree of $p \& q, p \vee q, \neg p$ etc. from the truth degrees of p, q (not discussing the question whether this is meaningful, i.e., $truth(p \& q)$ is *any* function of $truth(p)$, $truth(q)$. Let us admit that extensionality (truth-functionality) of fuzzy logic is not obviously counter-intuitive. Taking the most common truth-table definitions for $\&, \vee, \neg$ ($min, max, 1-x$), one should realize that it is not counterintuitive that $p \& \neg p$ has a positive truth-value (namely $min(truth(p), 1-truth(p))$) if $0 < truth(p) < 1$ – one can be young (in some degree) and at the same time non-young (in some degree). Thus truth degrees in fuzzy logic behave as truth-values in many valued logic. Our claim is that fuzzy logic should be understood as many-valued logic *sui generis* and should be confronted with classical systems of many-valued logic. (This has been done by several authors, see Section 4.)

If, on the other hand, p is a crisp proposition (i.e., can be only true or false) then, whatever our belief in p is (denote it $belief(p)$), $belief(p \& \neg p)$ should be *zero*; we are *sure* that $p \& \neg p$ is false, whether p is true or not. Now it is natural to assume that there are various possible situations (or possible worlds) and in each of them p is either true or false; $belief(p)$ is then understood as some measure (not necessarily probabilistic) of the set of all possible situations (worlds) in which p is true. This leads directly to some kind of modal logic with *Kripke semantics*, and clearly this modal logic is in general *not* extensional: $measure(p \& q)$ need not be a function of $measure(p)$, $measure(q)$, similarly for $p \vee q$. Our claim is that the logic of uncertainty, i.e., the logic of degrees of belief, is a (generalized) modal logic *sui generis* and should be confronted with classical systems of modal logic. (This has been done by various authors, see as well Section 6.)

Our last claim is that truth degrees should not be mistaken for degrees of belief and *vice versa*; still, it is possible to *combine* them; the natural home for this is *many-valued modal logic*. We shall discuss the details in Section 7.

3 Digression: Rule Based Systems

Let us first have a look at MYCIN-like rule-based systems (Shortliffe and Buchanan 1975), (Buchanan and Shortliffe 1984), (Duda et al. 1976)). They may be called classical or old-fashioned; even if still in use, they no longer seem to command theoretical interest. A rule base of such a system consists of *rules* of the form $A \to S(w)$ where A (antecedent) is typically an elementary conjunction of literals (e.g., p_1 & $\neg p_3$ & p_5), S (succedent) is a single literal (e.g., p_7) and w is weight of the rule (usually $\in [0,1]$ or $\in [-1,1]$). The system of rules must be loop-free in the obvious sense; leaf propositions (not occuring in any succedent) are *questions*. A *questionnaire* assigns to each question its weight (answer - degree of belief). *Combining functions* are truth tables for connectives, a function $CTR(a,w)$ computes the *contribution* of a rule with weight w provided that the global weight a of its antecedent, and a binary operation $\oplus$ that combines the contributions of rules; the *global weight* of a proposition p is the $\oplus$-sum of contributions of all rules leading to p.

Weights of rules are usually understood as *conditional beliefs*, sometimes even conditional probabilities; early criticism (Adams 1976, Johnson 1986) shows, however, that the interpretation in terms of conditional probabilities cannot be justified. A deeper and more detailed analysis was done by (Hájek and Valdes 1990, 1991), (Hájek 1988, 1989) (see also (Hájek et al. 1992: Chap. VI-VIII)) where an algebraic analysis (using the apparatus of ordered Abelian groups) is applied and a method of guarded use of these systems is suggested. But this method applies only to a particular subclass of rule based systems and guarantees only partial probabilistic soundness (using the apparatus of graphical probabilistic models). The main starting point is the observation (made also by others, e.g., Heckerman) that weights must not be identified with conditional belief but must be computed from them.

So far so good: the analysis using ordered Abelian groups and graphical models (hopefully) deserves interest, but what *logic* is involved here? The answer is: not much (except for highly interesting logical properties of ordered Abelian groups and, furthermore, the particular case when only three values (yes, no, unknown) are allowed.) Logically, these systems seem to be poorly motivated, they are truth-functional but clearly deal only with uncertainty, not with fuzziness (and should not be mistaken for rule-based systems based on fuzzy logic, see for example (Bonissone 1987). A more detailed analysis is to be found in the book (Hájek et al. 1992).

4 Many-Valued Logics

This section provides a quick survey of some basic systems of many-valued logics. We restrict ourselves to *many-valued propositional calculi*; thus our vocabulary consists of propositional variables $p_1, p_2, \ldots$, parentheses and some connectives (the choice of primitive connectives may vary from one system to another; but implication and negation are usually present in most cases).

It is generally recognized that pioneering work in many-valued logic was done by Lukasiewicz, through his discussion of three-valued logics (with the third value meaning "possible"), cf. (Lukasiewicz 1970). It is much less known and recognized that Gödel investigated extremely interesting systems of many-valued logic in connection with intuitionistic logic (Gödel 1993). (Rosser and Turquette 1952) is a classical monograph; an excellent recent and very informative monograph is (Gottwald 1988). The literature on *fuzzy logic* (started by Zadeh) is immense; a very up to date survey is found in (Dubois and Prade 1991a). An older paper of great importance, is (Pavelka 1979).

Our interest here is in logic in narrow sense of formal systems. We are now reviewing some concrete systems.

Semantics of Lukasiewicz's calculi.

Connectives are $\rightarrow$ and $\neg$; the truth functions are as follows:
$$impl(x, y) = 1 \text{ if } x \leq y,$$
$$impl(x, y) = 1 - x + y \text{ otherwise},$$
$$neg(x) = 1 - x.$$

From these, two different conjunctions and two different disjunctions are defined. Systems differ depending on the truth values allowed. There are finite-valued systems L_n ($n = 2, 3, 4, \ldots$) with the set of values $\{i/n \mid i = 0, 1, \ldots, (n-1)\}$; the countably infinite valued system L_ω has all rationals from $[0,1]$ for its truth values and L_∞ has the whole continuum $[0,1]$ of values.

Gödel's calculi.

Connectives are $\rightarrow, \neg, \&, \vee$; the semantics is:
$$impl(x, y) = 1 \text{ if } x \leq y,$$
$$impl(x, y) = y \text{ otherwise},$$
$$neg(0) = 1,$$
$$neg(x) = 0 \text{ otherwise};$$
$$conj(x, y) = min(x, y)$$
$$disj(x, y) = max(x, y).$$

The systems G_n, G_ω, G_∞ are defined analogously to the L-systems.

One should also mention an extensive literature on triangular norms and conorms as possible truth-function for conjunction and disjunction, and the induced truth-functions for implications. (Schweitzer and Sklar 1983) is a pioneering monograph on this subject; for detailed information see for example (Di Nola et al. 1989).

We now turn to matters of axiomatization. There are two main kinds of axiomatizations. The *first kind* concerns the axiomatization of 1-tautologies, i.e., of formulas having the value 1 under each evaluation of propositional variables. Some classical tautologies are 1-tautologies (under any of the discussed semantics), e.g., $(p \rightarrow p)$; but some are not, e.g. $p \vee \neg p$ in Gödel's semantics (note the relation to intuitionism!). The *second kind* of axiomatization deals with graded axioms and proofs; a graded proof is a sequence of pairs *(formula, truth-value)* satisfying some natural structural conditions; the existence of a graded proof with (A, α) as its conlusion means that A is provable (at least) to the degree α.

Gödel's calculi axiomatized.

(1-tautologies). The axioms of G_ω coincide with axioms of G_∞ and consist of the axioms of intuitionistic propositional logic (see for example (Kleene 1952)) plus the axiom of linearity:

$$(A \rightarrow B) \vee (B \rightarrow A).$$

The only deduction rule is *modus ponens*.

The axioms of G_n are those of G_ω plus a certain pigeon-hole axiom. These axiom systems are complete for 1-tautologies of the respective systems (cf. (Gottwald 1988)).

Lukasiewicz's calculi axiomatized.

L_3 was axiomatized by (Wajsberg 1931); the following is a complete axiomatization of 1-tautologies:

$$A \rightarrow (B \rightarrow A)$$

$$(A \rightarrow B) \rightarrow ((B \rightarrow C) \rightarrow (A \rightarrow C))$$

$$(\neg B \rightarrow \neg A) \rightarrow (A \rightarrow B)$$

$$((A \rightarrow \neg A) \rightarrow A) \rightarrow A$$

A similar complete axiomatization of 1-tautologies of L_ω (and of L_∞) was obtained by (Rose and Rosser 1958). Other Lukasiewicz systems have also axiomatizations of 1-tautologies, but they are more or less cumbersome. See (Gottwald 1988).

The system of Takeuti and Titani.

The system of Takeuti and Titani extends G_ω by Lukasiewicz's negation and some other connectives; it has about 50 axioms and provides a complete sequent calculus. The calculus is infinitary since it has an infinitary deduction rule

from all $A + \epsilon$ (ϵ dyadic) infer A

where $+$ is a new connective.

Calculi for graded truth: Pavelka's logic for L_ω, L_n.

L_ω and L_n are extended by names $\underline{\alpha}$ for each truth-value α. *Graded formulas* are pairs (A, α) where A is a formula and α is a truth-value. Among the logical axioms are

$(A, 1)$ for some natural A,
$(\underline{\alpha}, \alpha)$ for each α

There are *graded deduction rules*, such as a graded modus ponens:
from (A, α) and $(A \rightarrow B, \beta)$ infer $(B, \alpha \otimes \beta)$ where $\otimes$ is an appropriate function.

Each fuzzy set of formulas (i.e., mapping of the set of all formulas into truth-values) is a *fuzzy set of special axioms*; the definition of a graded proof from a fuzzy set F of special axioms is obvious. The accompanying *completeness theorem* says that the supremum of degrees of proof of A from F equals the infimum of values of A in evaluations V of formulas such that $F \subseteq V$, i.e., for each C, $F(C) \leq V(C)$.

Remarks

- The completeness proof makes heavy use of the fact that *impl* is continuous.

- We mention in passing (Mukaidono 1982), a paper dealing with fuzzy resolution.

- To claim that fuzzy logic should be understood primarily as many-valued logic, one has to analyze thesubstantiate literature on fuzzy logic from the many-valued point of view. This remains a task for the future.

5 Modal Logics

Modal logic became a part of mathematical logic through the advent of an adequeate formal semantics, namely the Kripke models. Basic systems deal with formulas built from propositional variables, logical connectives and the modality $\Box$ (the corresponding formation rule reads: if A is a formula then $\Box A$ is a formula). $\Box A$ is read "necessarily A"; the dual modality $\Diamond$ ("possibly") is defined as $\neg\Box\neg$. Other modalities are not excluded. In a general setting, a Kripke model is a triple $K = \langle W, \Vdash, S\rangle$, where W is a non-empty set of possible worlds, $\Vdash$ is a function assigning to each atomic formula P and each world w either 0 or 1 (truth evaluation) and S is some structure on W. In the classical systems, S is just a binary relation R on W, i.e., $R \subseteq W \times W$. Then $\Vdash$ extends to all formulas as follows (one writes $w \Vdash A$ for $\Vdash(A, w) = 1$):

$w \Vdash A \ \& \ B$ iff $w \Vdash A$ and $w \Vdash B$,

$w \Vdash \neg A$ iff not $w \Vdash A$,

$w \Vdash \Box A$ iff, for each v, such that wRv, $v \Vdash A$.

(Thus: "necessarily A" is true in w iff A is true in all worlds accessible from A)

The beauty of modal logic is based on the fact that simple assumptions on the properties of the accessibility relation R correspond to simple, complete axiom systems. We mention three examples:

Example 1. R is an equivalence relation (reflexive, symmetric, transitive). The corresponding axiom system is (S5). The axioms are

(0)   *propositional tautologies,*

(1) $\Box(A \rightarrow B) \rightarrow (\Box A \rightarrow \Box B)$,

(2) $\Box A \rightarrow \Box\Box A$,

(3) $\Box A \rightarrow A$,

(4) $\Diamond A \rightarrow \Box\Diamond A$.

Deduction rules are modus ponens and necessitation: "from A infer $\Box A$". (S5) is sound and complete with respect to all (finite) Kripke models whose accessibility relation is an equivalence relation. This means: (S5) proves A if and only if A is true in each world of each model $\langle K, \Vdash, R \rangle$ where R is an equivalence relation. (For a readable presentation see for example (Hughess and Cresswell 1968).)

Example 2. Provability logic (L). We have the axioms: (0), (1), (2) and Löb's axiom:

(5) $\Box(\Box A \to A) \to \Box A$.

This logic is sound and complete with respect to all finite Kripke models $K = \langle W, \Vdash, R \rangle$ such that R is a strict partial order of W with a unique root, so predecessors of each element are linearly ordered. Besides, (L) has a completely different completeness theorem, relating it to first-order arithmetic. We only sketch the result. Let PA be Peano arithmetic. An *arithmetical interpretation* of (L) is a mapping associating to each formula B of L a closed formula B^* of PA such that, for each C, D, $(C \& D)^*$ is $C^* \& D^*$, $(\neg C)^*$ is $\neg(C^*)$, $(\Box C)^*$ is $Pr(\overline{C^*})$ where Pr is the formalized provability predicate and $\overline{C^*}$ is the formal name of C^* (numeral expressing the Gödel number of C^*). Solovay's completeness theorem says that a formula A is provable in (L) iff for each arithmetical interpretation $*$, PA proves A^*. See (Smoryński 1985) for the details. (Note that arithmetically, Löb's axiom is a formulation of the celebrated Second Gödel's incompleteness theorem.)

Example 3. Tense logics: R is a strict order. We present an axiom system that is sound and complete for finite models $\langle W, \Vdash, R \rangle$ where R is a strict linear order of W. It is just the extension of (L) that results from adding the following axiom of linearity:

$$(\Diamond A \ \& \ \Diamond B) \to (\Diamond(A \ \& \ B) \lor \Diamond(A \ \& \ \Diamond B) \lor \Diamond(\Diamond A \ \& \ B)).$$

It this logic, $\Box A$ is to be read "always in the future", $\Diamond A$ "sometimes in the future"; thus the last axiom says "if both A and B will hold sometimes in the future, then either $A \ \& \ B$ will hold sometimes in the future or (sometimes in the future, A will hold and sometimes later B will hold) or conversely (B will hold, and sometimes later, A will hold)". The reader is assumed to agree that the fact that things seemingly so different as arithmetical provability and logic of future tense happen to be well related is an illustration of the beauty of modal logic. For treatment of tense logics see (Gabbay 1972), (Burghess 1984), and (Van Benthem 1991).

6 Uncertainty and Modal Logic

We shall now relate modal logic to uncertainty understood as degree of belief. To this end we shall assume Kripke models of the form $K = \langle W, \Vdash, S \rangle$, where S is (or defines) a measure defined for sets of worlds. To simplify the matter, we shall assume the set W of worlds to be finite and the measure to be defined for all subsets of W. On the other hand, we shall not necessarily assume a probabilistic measure. The degree of belief into A will be identified with the measure of the set $\{w \mid w \Vdash A\}$ of all worlds satisfying A. We shall consider three particular cases.

(1) $K = \langle W, \Vdash, P \rangle$ where P is a probability on W, i.e.,

$$P : W \to [0,1], \sum_{x \in W} P(x) = 1.$$

For $X \subseteq W$, $P(X) = \sum_{x \in X} P(x)$. This leads to *possibilistic logic*, cf. (Nilsson 1986).

(2) $K = \langle W, \Vdash, \pi \rangle$ where π is a possibility on W, i.e., $\pi : W \to [0,1]$, $max_{x \in W}\, \pi(x) = 1$.

For $X \subseteq W$, the possibility $\Pi(X) = max_{x \in X}\, \pi(x)$. This leads to *possibilistic logic*. Possibility has a dual notion – necessity: $N(X) = 1 - \Pi(W - X)$. (cf. Dubois and Prade 1988, 1990, 1991).

(3) $K = \langle W, \Vdash, m \rangle$ where m is a *basic belief assignment*, i.e., m maps the power set $P(W)$ into $[0,1]$ and $\sum_{X \subseteq W} m(X) = 1$. m is *regular* if $m(\emptyset) = 0$. The belief of X, $bel(X)$ is defined thus: $bel(X) = \sum_{Y \subseteq X} m(Y)$. Alternatively, one can define $bel'(X) = \sum_{\emptyset \neq Y \subseteq X} m(Y)$. These definitions coincide for regular assignments. This leads to the *Dempster-Shafer theory* and the logic of belief functions. See for example (Shafer 1976), (Smets et al. 1988), (Hájek and Harmanec 1992). The dual notion is plausibility, $pl(X) = 1 - bel(W - X)$.

It is known and easy to show that both probabilistic and possibilic models are particular cases of belief models; probabilistic models are in one-one correspondence with belief models such that the assignment $m(X)$ is positive only if X is singleton. And possibilistic models $\langle W, \Vdash, \pi \rangle$ have a one to one correspondence with belief models $\langle W, \Vdash, m \rangle$ such that the focal elements of m (i.e., sets X such that $m(X) \neq 0$) are nested, i.e., linearly ordered by inclusion. The correspondence is such that possibilities coincide with plausibilities given by nested belief models.

There is another very pleasing correspondence, described in (Ruspini 1986) and (Resconi et al., submitted). Consider (finite) Kripke models $\langle W, \Vdash, R, P \rangle$ where R is a reflexive binary relation on W and P is a probability on W. Use $\Box$ to define necessity and put $bel(A) = P(\Box A)$ (i.e., $P(\{w \mid w \Vdash \Box A\})$). Then bel is a regular belief function; each regular belief function is obtained from a Kripke model as above where R can be taken to be an equivalence. In other words, regular belief functions are in a one to one correspondence with probabilistic models of (S5).

As a last alternative, we would like to mention another suggestion by (Ruspini 1991): a *Ruspini model* has the form $\langle W, \|\!-, S \rangle$ where S is a fuzzy binary *similarity relation* on W, i.e., a mapping $S : W \times W \to [0, 1]$ satisfying some natural assumptions. Using S, Ruspini defines possibility distributions generalizing (2). For a detailed comparison of his approach with that of Dubois and Prade see (Esteva et al. 1993).

We now move to specific modalities capturing uncertainty.

First, we introduce unary modalities: take an $\alpha \in [0, 1]$ and define, for any model $K = \langle W, \|\!-, measure \rangle$, $K \|\!-\Box_\alpha A$ iff $measure(A) \geq \alpha$. (reed $\Box_\alpha A$ as "A is α -strongly believed". This is a well-behaved modality. Note that it was studied, for probabilistic measures and unary predicate logic in (Hájek and Havránek 1978) and (Hájek 1981). Other authors investigated other measures; notably, (Dubois and Prade 1990) investigated the resolution logic, based on the deduction rule

$$\frac{\Box_\alpha(A \vee B), \Box_\alpha(\neg B \vee C)}{\Box_\alpha(A \vee C)}$$

which is sound for necessity measures.

Second, we turn to binary modalities of *belief comparison*. Introduce a binary modality $\lhd$ and read a formula $A \lhd B$ as "B is at least as believed (probable, possible) as A". Formulas are built from propositional variables, connectives and the modality $\lhd$; in particular we have the formation rule if A, B are formulas then $A \lhd B$ is also a formula. This means that nested occurrences of $\lhd$ (e.g., $A \lhd (A \lhd B)$) are allowed. (Fariñas and Herzig 1991) introduced an axiom system QPL (qualitative possibilistic logic) which is sound for possibilistic models (i.e., models of the form $K = \langle W, \|\!-, \pi \rangle$ where π is a possibility distribution as above; $K \|\!- A \lhd B$ iff $\Pi(A) \leq \Pi(B)$). Their *axioms* are as follows:

(1) *tautologies*,
(2) $((A \lhd B) \text{ and } (B \lhd C)) \to (A \lhd C)$ (*transitivity*)
(3) $(A \lhd B) \vee (B \lhd A)$ (*linearity*)
(4) $\neg(true \lhd false)$
(5) $(false \lhd true)$
(6) $(A \lhd B) \to ((A \vee C) \lhd (B \vee C))$ (*monotonicity*

The *deduction rules* are modus ponens and a specific necessitation rule: from $A \to B$ infer $A \lhd B$. (Boutilier 1992) investigated QPL and related it to a system of tense logic with two unary modalities. Herzig showed (personal communication) that QPL$^+$ is axiomatized by QPL plus the axiom $(A \lhd B) \to \Box(A \lhd B)$, where $\Box C$ is $(\neg C) \lhd false$. (Bendová and Hájek 1993) showed that QPL is incomplete, presented its complete extension (by a slightly cumbersome axiom) and showed that the complete logic QPL* has a faithful interpretation in a tense logic with finite linearly preordered time (FLPOT); completeness of FLPOT was also proved. FLPOT has three basic necessity modalities G, H, I, meaning "in all future worlds, in all past worlds, in all present worlds". In FLPOT, $A \lhd B$ is interpreted as saying "in all worlds (past, present and future), if A holds then in some present or future world, B holds", in symbols: $\Box(A \to (JB \vee FB))$

where $\square C$ is HC & IC & GC, JC is $\neg I \neg C$, FC is $\neg G \neg C$. For details see (Bendová and Hájek 1993); the result is that QPL^+ is complete and can be understood as a sublogic of a certain tense logic, namely FLPOT with finite linearly preordered time.

A recent result of Hájek (paper in preparation) relates a variant of qualitative possibilistic logic to the *interpretability logic* developed and studied in connection with metamathematics of first-order arithmetic (see (Visser 1990) for a survey on interpretability logic).

These results show that qualitative possibilistic logics are well-behaved and closely related to established modal logics.

7 Vagueness and Uncertainty: Many Valued Modal Logics

We have observed that dealing with both vagueness and uncertainty naturally leads to many-valued modal logics. Such systems have been thoroughly investigated in the literature only quite recently; (Fitting 1992a, 1992b) are pioneering papers. Fitting's Kripke models are of the form $K = \langle W, \Vdash, S \rangle$ where $\Vdash$ maps *Atoms* × W into *Values* and S is a binary fuzzy relation on W (i.e., maps $W \times W$ into *Values*). This is related to Ruspini's similarity structures mentioned above. Fitting constructs a (Gentzen style) sequent calculus and proves its completeness (with respect to 1-tautologies).

One can consider many-valued possibilistic models $K = \langle W, \Vdash, S \rangle$ where $\Vdash$ is as above and S is a fuzzy subset of W, i.e., a possibility distribution. (Hájek and Harmancová 1993) investigate a logic with a binary modality $\lhd$ meaning *comparison of fuzzy truth values*: for each $i \in Values$, put $\tau_A(i) = \Pi(\{w \mid \parallel A \parallel_w \geq i\})$ (possibility of A having the value of at least i); then τ_A is a fuzzy truth value (fuzzy subset of *Values*) and we define $K \Vdash A \lhd B$ iff $(\forall i)(\tau_A(i) \leq \tau_B(i))$. This modality behaves reasonably and is related, in (Hájek and Harmancová 1993), to a many valued temporal logic with linearly preordered time. Further systems of qualitative fuzzy possibilistic logic are conceivable (and of cause, qualitative fuzzy probabilistic or Dempster-Shafer logic); this is a task for future research.

8 Conclusion

We have tried to show that the proper logical home for logics of approximate reasoning is many-valued modal logic; better, the family of various many-valued modal logics. Many-valuedness relates to fuzziness; fuzziness is (mostly, and correctly) extensional (truth-functional). Uncertainty as degree of belief (probability, possibility, Dempster-Shafer belief function) relates to modalities, possible worlds and structures on possible worlds. Various logical systems have been investigated, showed to be complete for suitable semantics, and related to established logical systems, e.g., of tense logic. As a consequence, that this direction of research appears promising: several other logics of approximate reasoning can be defined and investigated, in particular, probabilistic and Dempster-Shafer logics, crisp or fuzzy. This being done, one should try to synthetize and isolate the most important systems. The goal is to get a unified logical approach to approximate reasoning.

References

Adams, J.B.: "A Probability Model of Medical Reasoning and the MYCIN Model," in: *Mathematical Biosciences*, Vol.32 (1976) 177-186.

Bendová, K., and P. Hájek: "Possibilistic logic as tense logic," in: Piera Carreté, N. et al. (eds.), *Qualitative Reasoning and Decision Technologies*, CIMNE Barcelona: Proceedings QUARDET'93 (1993) 441-450.

Bonissone, P.P.: "Summarizing and Propagating Uncertain Information with Triangular Norms," *International Journal Approximate Reasoning* 1 (1987) 71-101.

Boutilier, C.: "Modal logics for qualitative possibility and beliefs," in: Dubois, D. et al. (eds.), *Uncertainty in Artificial Intelligence* VIII, San Mateo, CA: Morgan-Kaufmann Publishers (1992) 17-24.

Buchanan, B.G., and E.H. Shortliffe: "Rule Based Expert Systems," in: *MYCIN Experiments of the Stanford Heuristic Programming Project*, Reading, MA: Addison-Wesley Pub.Comp. (1984) 739.

Burghess, J.P.: "Basic tense logics," in: Gabbay, D. and F. Guenthner (eds.), *Handbook of Philosophical Logic*, Vol.II. Reidel, 1984.

Di Nola, A., S. Sessa, W. Pedrycz, and E. Sanchez: *Fuzzy Relation Equations and Their Applications to Knowledge Engineering.* Dordrecht: Kluwer Academic Publishers, 1989.

Dubois, D., and H. Prade: "An Introduction to Possibilistic and Fuzzy Logics," in: Smets et al., *Non-Standard Logics for Automated Reasoning.* London: Academic Press (1988), p.287-326.

Dubois, D., and H. Prade: "Resolution Principles in Possibilistic Logic," in: *International Journal of Approximate Reasoning*, Vol.4. No.1 (1990) 1-21.

Dubois, D., and H. Prade: "Fuzzy sets in approximate reasoning, Part 1: Inference with possibility distribution," *Fuzzy Sets and Systems*, No.40 (1991a) 143-202.

Dubois, D., and H. Prade: "Fuzzy sets in approximate reasoning, Part 2: Logical approaches," in: *Fuzzy Sets and Systems*, No.40 (1991b) 203-244.

Duda, R.O., P.E. Hart, and N.J. Nilsson: "Subjective Bayesian Methods for Rule-Based Inference Systems," in: *Proceedings Nat. Comp .Conf.*, AFIPS, (1976) 1075-1082.

Esteva, F., P. García, and L. Godo: "On the relationship between preference and similarity based approaches to possibilistic reasoning," in: *Proceedings FUZZ-IEEE'93*, 1993.

Fariñas del Cerro, L., and A. Herzig: "A modal analysis of possibilistic logic," in: Kruse et al. (eds.), *Symbolic and Quantitative Approaches to Uncertainty.* LNCS 548, Springer-Verlag (1991) 58 ff. Also in: Jorrand and Kelemen (eds.), *Fundamentals of AI research*, Lecture Notes in AI 535, Springer-Verlag (1991) 11 ff.

Fitting M.: "Many-valued modal logics I," in: *Fundamenta Informatical* 15 (1992a) 235-254.

Fitting, M.: "Many-valued modal logics II," in: *Fundamenta Informatical* 17 (1992) 55-73.

Gabbay, D.M.: "Tense logics with discrete moments of time I," in: *Journal of Philosophical Logic* 1 (1972) 35 ff.

Gödel, K.: Zum intuitionistischen Aussugenkalkül, Ergebnisse eines Math. Colloquium 4 (1993) 40 ff. (see also Gödel's collected works Vol.1)

Godo L., and R. Lopez de Mantaras: "Fuzzy logic," in: *Encyclopedia of Computer Science*, to appear.

Gottwald, S.: *Mehrwertige Logik*. Berlin: Akademie-Verlag, 1988.

Hájek, P.: "Decision Problems of Some Statistically Motivated Monadic Modal Calculi," in: *International Journal Man-Machine Studies*, Vol.15 (1981) 351-358.

Hájek, P.: "Towards a Probabilistic Analysis of MYCIN-like Expert Systems," in: *COMPSTAT'88 Copenhagen*. Heidelberg: Physica-Verlag, 1988.

Hájek, P.: "Towards a probabilistic analysis of MYCIN-like Systems II," in: Plander (ed.), *Artificial Intelligence and Information-Control Systems of Robots*. Amsterdam: North Holland, 1989.

Hájek, P., and D. Harmancová: "A comparative fuzzy modal logic," in: Klement, Slany (eds.), *Fuzzy Logic in Artificial Intelligence*. Lecture Notes in AI 695, Springer-Verlag (1993) 27-34.

Hájek, P., and D. Harmanec: "An exercise in Dempster-Shafer theory," in: *International Journal General Systems* 20 (1992) 137-140.

Hájek, P., and T. Havránek: *Mechanizing Hypothesis Formation: Mathematical foundations for a general theory*. Berlin, Heidelberg: Springer-Verlag (1978) 396.

Hájek, P., T. Havránek, and R. Jirousek: *Uncertain Information Processing in Expert Systems*. Boca Raton: CRC Press (1992) 285.

Hájek, P., and J. Valdes: "Algebraic Foundations of Uncertainty Processing in Rule-based Expert Systems (group-theoretical approach)," in: *Computers and Artificial Intelligence* Vol.9 (1990) 325-344.

Hájek, P., and J. Valdes: "Generalized Algebraic Foundations of Uncertainty Processing in Rule-based Expert Systems (dempsteroids)," in: *Computers and Artificial Intelligence* Vol.10 (1991) 29-42.

Hughess, C.E., and M.J. Cresswell: *An Introduction to Modal Logic*. London: Methuen, 1968.

Johnson, R.W.: "Independence and Bayesian Updating Methods," in: *Uncertainty in AI*, Amsterdam: North Holland (1986) 197.

Kleene, S.C.: *Introduction to Metamathematics*. Amsterdam: North Holland, 1952.

Lukasiewicz, J.: *Selected Works*. Amsterdam: North Holland, 1970.

Mukaidono, M.:" Fuzzy inference in resolution style," in: Yager (ed.), *Fuzzy sets and possibility theory-recent developments*. Peyamon Press (1982) 224-231.

Nilsson, N.J.: "Probabilistic logic," in: *Artificial Intelligence* 28 (1986) 71-87.

Pavelka, J.: "On fuzzy logic I - Many valued rules of inference," in: *Zeitschr. f. Math. Logic und Grundlagen d. Math.* 25 (1979a) 45-52.

Pavelka, J.: "On fuzzy logic II - Enriched residuated lattices and semantics of propositional calculi," in: *Zeitschrift f. Math. Logik und Grundlagen d. Math.* 25 (1979b) 119-134.

Pavelka, J.: "On fuzzy logic III - Semantical completeness of some many-valued propositional calculi," in: *Zeitschrift f. Math. Logic und Grundlagen d. Math.* 25 (1979c) 446-464.

Resconi, G., G.J. Klir, U. St.Clair, and D. Harmanec: "On the integration of uncertainty theories," in: *International J. of Uncertainty, Fuzziness, and Knowledge-Based Systems* 1, 1993.

Resconi, G., G.J. Klir, and U. St.Clair: "Hierarchical uncertainty metatheory based upon modal logic," in: *Int. J. General Systems* 21 (1992) 23-50.

Rose, A., and J.B. Rosser: "Fragments of many-valued statement calculi," in: *Trans. Amer. Math. Soc.* 87 (1958) 1-53.

Rosser, J.B., and A.R. Turquette: *Many-valued logic.* Amsterdam: North Holland, 1952.

Ruspini, E.H.: *The logical foundations of evidential reasoning.* Techn. Note 408, AI Center. Menlo Park CA: SRI International, 1986 (revised 1987).

Ruspini, E.H.: "On the semantics of fuzzy logic," in: *Int. J. Approach Reasoning* 5 (1991) 45-88.

Schweitzer, B., and A. Sklar: *Probability metric spaces.* New York: North Holland, 1983.

Shafer, G.: '*A Mathematical Theory of Evidence.* Princenton: Princenton University Press, 1976.

Shortliffe, E.H., and B.G. Buchanan: "A Model of Inexact Reasoning in Medicine," in: *Mathematical Biosciences* Vol.23 (1975) 351-379.

Smets, P., A. Mamdani, D. Dubois, and H. Prade (eds.): *Non-standard Logics for Automated Reasoning.* London: Academic Press, 1988.

Smoryński, C.: *Self-reference and Modal Logic.* Springer Verlag, 1985.

Takeuti, G., and S. Titani: "Fuzzy logic and fuzzy set theory," in: *Archive for Math. Logic*, to appear.

Van Benthem, J.: *The Logic of Time.* Dordrecht: Kluwer Academic Publisher, 1991.

Visser, A.: "Interpretability Logic," in: *Mathematical Logic, Proceedings of Heyting Conference Bulgaria 1988.* Plenum Press, 1990.

Wajsberg, M.: "Axiomatization of the three-valued propositional calculus," (Polish with German summary) in: *C.R. Soc. Sci. Lett Varsovie*, Cl. 3, Vol. 24 (1931) 269-283. (Translation in: M. Wajsberg, *Logical Works.* Warsaw: Polish Academy of Sciences, 1977.

Gentzen Sequent Calculus for Possibilistic Reasoning

Churn Jung Liau[1] and Bertrand I-Peng Lin[2]

[1] Institute of Information Science, Academia Sinica,
Taipei, Taiwan, ROC, E-mail: liau@iis.sinica.edu.tw
[2] Department of Computer Science and Information Engineering
National Taiwan University, Taipei, Taiwan, ROC

Abstract. Possibilistic logic is an important uncertainty reasoning mechanism based on Zadeh's possibility theory and classical logic. Its inference rules are derived from the classical resolution rule by attaching possibility or necessity weights to ordinary clauses. However, since not all possibility-valued formulae can be converted into equivalent possibilistic clauses, these inference rules are somewhat restricted. In this paper, we develop Gentzen sequent calculus for possibilistic reasoning to lift this restriction. This is done by first formulating possibilistic reasoning as a kind of modal logic. Then the Gentzen method for modal logics generalized to cover possibilistic logic. Finally, some properties of possibilistic logic, such as Craig's interpolation lemma and Beth's definability theorem are discussed in the context of Gentzen methods.

1 Introduction

The literature on uncertainty reasoning mechanisms based on Zadeh's possibility theory (Zadeh 1978) has grown rapidly. (See (Dubois et al. 1990) for a survey). Possibilistic logic (Dubois and Prade 1988) is one of the most important approaches in this direction. However, since the deduction method of possibilistic logic is based on the classical resolution rule (Robinson 1965), it is restricted to formulae in clausal form. On the other hand, it has been pointed out that possibilistic logic can be viewed as a generalization of modal logics (Dubois and Prade 1988, Dubois et al. 1988, Farinas del Cerro and Herzig 1991, Liau and Lin 1992), so it should be possible to use the deduction method of modal logics to do possibilistic reasoning. In this paper, we will investigate an implementation of this idea to lift the clausal form restriction of possibilistic resolution. To do this, we first formulate possibilistic reasoning as a kind of modal logic with multi modal operators. Then, we generalize the sequent calculus of normal modal logics developed by (Fitting 1983) to our logic. Finally, some logical properties of our system are discussed.

1.1 Possibilistic resolution rule

In possibilistic logic, two certainty measures, called possibility and necessity measures respectively, are assigned to the well-formed formulae of classical propositional logic. Specifically, let Π and N denote measures for possibility and necessity respectively, then the following laws must be satisfied for all well-formed formulae f and g:

(i) $\Pi(f)$ and $N(f) \in [0,1]$
(ii) $N(f) = 1 - \Pi(\neg f)$,
(iii) $\Pi(\top) = 1$ and $\Pi(\bot) = 0$, and
(iv) $N(f \wedge g) = \min(N(f), N(g))$.

Thus, if f is a well-formed formula (wff) of propositional logic, then $(f(Nc))$ and $(f(\Pi c))$, where $c \in [0,1]$, are wffs of possibilistic logic. When f is a classical clause, the wffs are called possibilistic clauses. The intended meanings of the wffs are $N(f) \geq c$ and $\Pi(f) \geq c$ respectively. The resolution rule of possibilistic logic is defined as follows:

$$\frac{(f \ w_1) \quad (g \ w_2)}{(R(f,g) \ w_1 * w_2)}$$

where $R(f,g)$ is the classical resolvent of f and g, and $*$ is defined by

$$(N \ c) * (N \ d) = (N \ min(c,d))$$
$$(N \ c) * (\Pi \ d) = \begin{cases} (\Pi \ d) \ if \ c+d > 1 \\ (\Pi \ 0) \ if \ c+d \leq 1 \end{cases}$$
$$(\Pi \ c) * (\Pi \ d) = (\Pi \ 0)$$

By the axiom (iv) above, any wff of the form $(f(Nc))$ can be converted into an equivalent set of possibilistic clauses, and if only the necessity measure N is involved, the refutational completeness of this resolution rule can be proved (Dubois et al. 1990). However, if possibility-valued wffs are involved, then the completeness result does not hold in general. The following simple example illustrates this point.

Example 1. Consider the assumption set $B = \{(p \wedge q(\Pi 0.7)), (p \wedge q \supset r \ (N \ 0.6))\}$ and the wff $f = (r \ (\Pi \ 0.7))$ in possibilistic logic. Then obviously, f should be derivable from B.[3] However, since $(p \wedge q \ (\Pi \ 0.7))$ is not a possibilistic clause, the resolution rule cannot be applied. A possible approach to overcome the difficulty may be to infer $(p \ (\Pi \ 0.7))$ and $(q \ (\Pi \ 0.7))$ firstly. This indeed helps some situations if the second sentence is either $(p \supset r \ (N \ 0.6))$ or $(q \supset r \ (N \ 0.6))$, but it is useless in the present case.

To remove the restriction of clausal form, we must consider more general deduction methods. Gentzen sequent calculus is one possible choice. It was first developed by Gentzen for classical and intuitionistic logic (Gentzen 1969), and then extended to modal logics by (Fitting 1983). We try to provide a suitable Gentzen system for possibilistic reasoning in the following sections. But before doing this, we must reformulate possibilistic logic as a general modal logic.

---

[3] Replacing "$p \wedge q$" with a new propositional letter and using the given resolution rule will justify this.

2 Quantitative Modal Logic

Possibilistic reasoning is first formulated as modal logic in (Farinas del Cerro and Herzig 1991). In (Liau and Lin 1992), we call the resultant logic quantitative modal logic (QML) and investigated axiomatic systems for it. We review QML's syntax and semantics briefly.

2.1 Syntax

QML is an extension of propositional logic with four classes of quantitative modal operators: $\langle c \rangle$, $\langle c \rangle^+$, $[c]$ and $[c]^+$ for all $c \in [0,1]$.

In addition to the syntactic rules for propositional logic, the following one is added to the formation rules of QML:

if f is a wff, so are $\langle c \rangle f$, $\langle c \rangle^+ f$, $[c]f$ and $[c]^+ f$ for all $c \in [0,1]$.

The intended meaning of $\langle c \rangle f$, $\langle c \rangle^+ f$, $[c]f$ and $[c]^+ f$ is $\Pi(f) \geq c$, $\Pi(f) > c$, $N(f) \geq c$ and $N(f) > c$ respectively.

We usually use lower case letters (sometimes with indices) p, q, r to denote atomic formulae and f, g, h to denote wffs. We also assume that all classical logical connectives are available, either as primitives or as abbreviations.

Note that we have enhanced the expressive power of possibilistic logic so that we can represent more complex wffs in QML. For example, we can represent higher order uncertainty in a wff like $\neg[0.7]^+(p \supset \langle 0.8 \rangle q)$. The importance of higher order uncertainty is highlighted in (Gaifman 1986). Furthermore, the syntax makes it easier to combine QML and other intensional logics. For instance, we can use $B\langle 0.6 \rangle p$ and $\langle 0.6 \rangle Bp$ to represent "the agent believes that the possibility of p is at least 0.6" and "the possibility of the agent believing p is at least 0.6" respectively.

Following an idea in (Fitting 1983), we may classify the non-literal wffs of QML into six categories according to the formulae's main connective. We list the alternatives in Fig. 1. We will abuse the notation and use $\alpha, \alpha_1, \cdots$, etc. to denote wffs of the respective types and their immediate subformulae.

2.2 Semantics

We now turn to the semantics of QML. Define a *possibility frame* $F = \langle W, R \rangle$, where W is a set of possible words and $R : W^2 \rightarrow [0,1]$ is a fuzzy accessibility relation on W. Let PV and FA denote the set of all propositional variables and the set of all wffs respectively. Then a model of QML is a triple $M = \langle W, R, TA \rangle$, where $\langle W, R \rangle$ is a possibility frame and $TA : W \times PV \rightarrow \{0,1\}$ is a truth value assignment for all worlds. A proposition p is said to be true at a world w iff $TA(w, p) = 1$. Given R, we can define a possibility distribution R_w for each $w \in W$ such that $R_w(s) = R(w, s)$ for all s in W. Similarly, we can define TA_w for each $w \in W$ such that $TA_w(p) = TA(w, p)$ for all p in PV. Thus, a model can be equivalently written as $\langle W, \langle R_w, TA_w \rangle_{w \in W} \rangle$. Given a model $M = \langle W, R, TA \rangle$, α wffs and their component formulae

α	α_1	α_2
$f \wedge g$	f	g
$\neg(f \vee g)$	$\neg f$	$\neg g$
$\neg(f \supset g)$	f	$\neg g$
$\neg\neg f$	f	f

β wffs and their component formalae

β	β_1	β_2
$f \vee g$	f	g
$\neg(f \wedge g)$	$\neg f$	$\neg g$
$f \supset g$	$\neg f$	g

ν and ν^+ wffs (with parameter c) and their component formulae

$\nu(c)$	$\nu_0(c)$	$\nu^+(c)$	$\nu_0^+(c)$
$[c]f$	f	$[c]^+f$	f
$\neg\langle 1-c\rangle^+ f$	$\neg f$	$\neg\langle 1-c\rangle f$	$\neg f$

π and π^+ wffs (with parameter c) and their component formulae

$\pi(c)$	$\pi_0(c)$	$\pi^+(c)$	$\pi_0^+(c)$
$\langle c\rangle f$	f	$\langle c\rangle^+ f$	f
$\neg[1-c]^+ f$	$\neg f$	$\neg[1-c]f$	$\neg f$

Fig. 1. Classification of QML non-literal wffs

we can define the satisfaction relation $\models_M\,\subseteq W \times FA$ as follows. Define $N_w(f) = \inf\{1 - R(w, s) \mid s \models_M \neg f, s \in W\}$. N_w is just the necessity measure induced by the possibility distribution R_w as defined in (Dubois and Prade 1988). Then, $w \models_M [c]f$ (resp. $w \models_M [c]^+f$) iff $N_w(f) \geq c$ (resp. $> c$). The satisfaction relations for $\langle c\rangle f$ and $\langle c\rangle^+ f$ are defined analogously by replacing $N_w(f)$ with $\Pi_w(f) = 1 - N_w(\neg f)$. For convenience, we define $\sup \emptyset = 0$ and $\inf \emptyset = 1$. Furthermore, the satisfaction of all other wffs is defined as usual in classical logic.

A wff f is said to be valid in $M = \langle W, R, TA\rangle$, write $\models_M f$, iff for all $w \in W, w \models_M f$. If S is a set of wffs, then $\models_M S$ means that for all $f \in S$, $\models_M f$. If $\mathbf{C}$ is a class of models and S is a set of wffs, then we write $S \models_{\mathbf{C}} f$ to say that for all $M \in \mathbf{C}$, $\models_M S$ implies $\models_M f$. Note that "$\models_{\mathbf{C}}$" is defined "model by model". It can also be defined "world by world", i.e., for all $M \in \mathbf{C}$ and w of M, $w \models_M S$ implies $w \models_M f$. However, when S is finite, the alternative definition is equivalent to $\models_{\mathbf{C}} \bigwedge S \supset f$, so we would not consider it separately. A model is called serial iff $\forall w \in W, \sup_{s \in W} R(w, s) = 1$. As pointed out in (Liau and Lin 1992), the consequence relation in serial models exactly reflects possibilistic reasoning, so in what follows, we will mainly focus on the development of deduction rules, and $S \models_{\mathbf{C}} f$ is abbreviated as $S \models f$ if $\mathbf{C}$ is the class of all serial models.

3 Gentzen Sequent Calculus

The Gentzen sequent calculus for possibilistic reasoning is a generalization of that for modal logics (Fitting 1983), and is an extension of classical logic. Therefore the system will contain all rules for classical logic. The principle of additional rules for possibilistic reasoning depends on the truth of wffs in different possible worlds. For instance, if $M = \langle W, R, TA \rangle$ is a model, and S is a set of wffs true in a world $w \in W$, then for all worlds w' such that $R(w, w') \geq c$, what wffs should be true in w'? Obviously, any wff f such that $N_w(f) > 1 - c$ or $\Pi_w(\neg f) < c$ should satisfy the requirement. Thus, in general, we have the following definition:

Definition 1. Define the world-alternative function K and the strict world-alternative function K^+ as follows.

$$K, K^+ : 2^{FA} \times [0, 1] \to 2^{FA},$$

$$K(S, c) = \{\nu_0(c') \mid c' \geq c, \nu(c') \in S\} \cup \{\nu_0^+(c') \mid c' \geq c, \nu^+(c') \in S\}$$

and

$$K^+(S, c) = \{\nu_0(c') \mid c' > c, \nu(c') \in S\} \cup \{\nu_0^+(c') \mid c' \geq c, \nu^+(c') \in S\}.$$

Intuitively, $K(S, c)$ (resp. $K^+(S, c)$) contains all wffs f such that $N(f) \geq c$ (resp. $N(f) > c$) is derivable from S by inequality constraints. Thus, it is readily verified that if S is true in w and $R(w, w') \geq c$ (resp. $> c$), then $K^+(S, 1 - c)$ (resp. $K(S, 1 - c)$) is true in w'.

Dually, we have the following definition for the falsity of a set S of wffs in a world w.

Definition 2. Define the dual world-alternative function $\hat{K}$ and strict dual world alternative function $\hat{K}^+$ as follows.

$$\hat{K}, \hat{K}^+ : 2^{FA} \times [0, 1] \to 2^{FA},$$

$$\hat{K}(S, c) = \{\pi_0(c') \mid c' \leq 1 - c, \pi(c') \in S\} \cup \{\pi_0^+(c') \mid c' \leq 1 - c, \pi^+(c') \in S\}$$

and

$$\hat{K}^+(S, c) = \{\pi_0(c') \mid c' \leq 1 - c, \pi(c') \in S\} \cup \{\pi_0^+(c') \mid c' < 1 - c, \pi^+(c') \in S\}.$$

Similarly, we can verify that if S is false in w and $R(w, w') \geq c$ (resp. $> c$), then $\hat{K}^+(S, 1-c)$ (resp. $\hat{K}(S, 1-c)$) is false in w'.

We are now ready to present the Gentzen deduction system for possibilistic reasoning. First, let us introduce the notation of a sequent. Let U, V be two sets of wffs, and let $U \longrightarrow V$ denote a sequent. The intuitive meaning of the sequent $U \longrightarrow V$ is that if all elements of U are true, then at least one of the wffs in V is true.

We present the Gentzen sequent calculus for QML in Fig. 2 by using the notation introduced in Fig. 1. We use "U, f" to denote the union of a set of wffs U and a wff f. Every rule consists of one or two upper sequents called *premises* and of a lower sequent called *conclusion*. The schema f stands for any literals.

1. Axioms:

$$U, f \longrightarrow V, f \qquad U, \bot \longrightarrow V \qquad U \longrightarrow V, \top \qquad U \longrightarrow V, \pi(0)$$
$$U, f, \neg f \longrightarrow V \qquad U, \nu^+(1) \longrightarrow V \qquad U \longrightarrow V, \nu(0) \qquad U \longrightarrow V, f, \neg f$$
$$U \longrightarrow V, \neg \bot \qquad U, \pi^+(1) \longrightarrow V \qquad U, \neg \top \longrightarrow V$$

2. Classical rules:

$$\alpha L: \quad \frac{U, \alpha_1, \alpha_2 \longrightarrow V}{U, \alpha \longrightarrow V} \qquad\qquad \beta R: \quad \frac{U \longrightarrow V, \beta_1, \beta_2}{U \longrightarrow V, \beta}$$

$$\beta L: \quad \frac{U, \beta_1 \longrightarrow V \quad U, \beta_2 \longrightarrow V}{U, \beta \longrightarrow V} \qquad \alpha R: \quad \frac{U \longrightarrow V, \alpha_1 \quad U \longrightarrow V, \alpha_2}{U \longrightarrow V, \alpha}$$

$$\neg^2 L: \quad \frac{U, \neg\neg f \longrightarrow V}{U, f \longrightarrow V} \qquad\qquad \neg^2 R: \quad \frac{U \longrightarrow V, \neg\neg f}{U \longrightarrow V, f}$$

3. Common rules for possibilistic reasoning:

$$\pi L_1: \text{if } c > 0 \quad \frac{K^+(U, 1-c), \pi_0(c) \longrightarrow \hat{K}^+(V, 1-c)}{U, \pi(c) \longrightarrow V}$$

$$\pi L_2: \quad \frac{K(U, 1-c), \pi_0^+(c) \longrightarrow \hat{K}(V, 1-c)}{U, \pi^+(c) \longrightarrow V}$$

$$\nu R_1: \text{if } c < 1 \quad \frac{K^+(U, c) \longrightarrow \hat{K}^+(V, c), \nu_0^+(c)}{U \longrightarrow V, \nu^+(c)}$$

$$\nu R_2: \quad \frac{K(U, c) \longrightarrow \hat{K}(V, c), \nu_0(c)}{U \longrightarrow V, \nu(c)}$$

4. Rule D (for the serality of models):

$$\frac{K^+(U, 0) \longrightarrow \hat{K}^+(V, 0)}{U \longrightarrow V}$$

Fig. 2. Gentzen Sequent Calculus for QML

We define the *global assumption rule* for a set of wffs B as follows: if $f \in B$, then

$$\frac{U \longrightarrow V, \neg f}{U \longrightarrow V} \qquad\qquad \frac{U, f \longrightarrow V}{U \longrightarrow V}$$

3.1 Soundness and completeness

Next, we will consider the soundness and completeness of Gentzen-type proof methods. First, we define the truth of a sequent $U \longrightarrow V$ in a world. Let $M = \langle W, R, TA \rangle$ be a model and $w \in W$, then $w \models_M U \longrightarrow V$ iff $w \models_M U$ implies there is at least a wff f in V such that $w \models_M f$. Then the validity of a sequent in a model and in a class of models can be defined analogously to that of a wff, and the notations $\models_M U \longrightarrow V$, $\models U \longrightarrow V$, and $B \models U \longrightarrow V$ can also be used for sequents, where B is a set of wffs. We also use the convention that $\longrightarrow V$ and $U \longrightarrow$ denote $\emptyset \longrightarrow V$ and $U \longrightarrow \emptyset$, respectively. A proof of a sequent $U \longrightarrow V$ from a set of wffs B is a sequence of sequents $\varphi_1, \varphi_2, \cdots, \varphi_n$ such that for each i, either φ_i is the instance of some axiom of the Gentzen system, or φ_i is the result of applying the rules in Fig. 2 or the global assumption rule to the earlier sequents of this sequence. We write $B \vdash (U \longrightarrow V)$ to mean that there exists a proof of $U \longrightarrow V$ from B.

To give the soundness theorem of the Gentzen system, we need another definition. A sequent $U \longrightarrow V$ is called *short* iff $U \cup V$ is finite. We note that a short sequent $U \longrightarrow V$ is true in a world iff the wff "$\bigwedge U \supset \bigvee V$" is true in that world, where $\bigwedge U$ and $\bigvee V$ denote the conjunction of U and the disjunction of V respectively.

Theorem 3 (Soundness). *If B is a finite set of wffs and $U \longrightarrow V$ is a short sequent, then $B \vdash (U \longrightarrow V)$ implies $B \models (U \longrightarrow V)$*

Proof. All axioms are valid (Liau and Lin 1992). As to the inference rules, we consider πL_1 as an example. If its conclusion is false in some world w, then $w \models \pi(c)$. This guarantees the existence of another world w' such that $R(w, w') \geq c$ if the set of possible worlds W is finite[4], so by the remarks following Definition 1 and 2, the premise is false in w'.

Usually, we want to prove only a wff f from a finite global assumption set B, so it is useful to state a restricted version of the soundness theorem if we identify a wff f with a sequent $\longrightarrow \{f\}$.

Corollary 4. $B \vdash f$ *implies* $B \models f$

Completeness can be proven by the model existence theorem for QML (Liau and Lin 1992). However, since we use unsigned formulae as opposed to the signed version in (Fitting 1983), some tedious steps are needed. In particular, the rules $\neg^2 L$ and $\neg^2 R$ are needed for completeness. We believe that the two rules can be removed by using signed formulae without affecting completeness.

---

[4] By a corollary in (Liau and Lin 1992), i.e., the finite model property, we can assume that W is finite without loss of generality

Theorem 5 (Completeness). *1. Let $U \longrightarrow V$ be a short sequent, then $B \models$*
$U \longrightarrow V$ implies $B \vdash U \longrightarrow V$.
2. $B \models f$ iff $B \vdash f$

We return to Example 1 to see how our Gentzen system can derive the expected result. The deduction is shown in Fig. 3, and the rules used from above are βL, πL_1 with $c = 0.7$, and with the global assumption rule twice used. Since the right one of the two topmost sequents is an instance of an axiom, and the other one can be derived from the axioms by classical rules, this is indeed a proof in Gentzen system.

$$\frac{\frac{\dfrac{\neg(p \wedge q), p \wedge q \longrightarrow r \qquad\qquad\qquad\qquad\qquad r, p \wedge q \longrightarrow r}{p \wedge q \supset r, p \wedge q \longrightarrow r}}{[0.6](p \wedge q \supset r), \langle 0.7 \rangle (p \wedge q) \longrightarrow \langle 0.7 \rangle r}}{\dfrac{[0.6](p \wedge q \supset r) \longrightarrow \langle 0.7 \rangle r}{\langle 0.7 \rangle r}}$$

Fig. 3. A Gentzen deduction of Example 1

4   Interpolation Lemma and Definability Theorem

In the sequel, we explore some logical properties of QML by exploiting the soundness and completeness of the Gentzen system.

First, we discuss the interpolation lemma. The interpolation lemma of classical logic is due to Craig (Craig 1957a, Craig 1957b). Here, it is be generalized to QML.

Definition 6. If $U \longrightarrow V$ is a sequent, then f is an interpolation formula for $U \longrightarrow V$ iff

(i) all propositional variables of f are common to U and V. In other words, $Pvar(f) \subseteq Pvar(U) \cap Pvar(V)$, where $Pvar(S) = \bigcup_{g \in S} Pvar(g)$ and $Pvar(g)$ are the propositional variables occurring in g.
(ii) $\models U \longrightarrow f$ and $\models f \longrightarrow V$.

We define an interpolation formula for $f \supset g$ as one for $f \longrightarrow g$, where f and g are two wffs. We essentially follow Fitting's method (Fitting 1983) to prove the interpolation lemma.

Lemma 7 (Craig's Interpolation Lemma). *If $\models f \supset g$, then $f \supset g$ has an interpolation formula.*

Proof. By the completeness of Gentzen system, $\models f \supset g$ implies $\models f \longrightarrow g$, and in turn implies $\vdash f \longrightarrow g$. We will try to show that

1. all axioms of the Gentzen system have the Craig property, and
2. for all rules in the Gentzen system, if the premise(s) has the Craig property, then the conclusion has, too.

Then, $f \longrightarrow g$ has an interpolation formula, and this is also the interpolation formula of $f \supset g$ by definition.

We use $U \xrightarrow{h} V$ to denote that h is an interpolation formula of $U \longrightarrow V$.

First, we show that all axioms have interpolation formulae by the following list.

$$U, f \xrightarrow{f} V, f \qquad U, \bot \xrightarrow{\bot} V \qquad U \xrightarrow{\top} V, \top \qquad U \xrightarrow{\top} V, \pi(0)$$

$$U, f, \neg f \xrightarrow{\bot} V \quad U, \nu^+(1) \xrightarrow{\bot} V \quad U \xrightarrow{\top} V, \nu(0) \quad U \xrightarrow{\top} V, f, \neg f$$

$$U \xrightarrow{\top} V, \neg \bot \quad U, \pi^+(1) \xrightarrow{\bot} V \quad U, \neg \top \xrightarrow{\bot} V$$

Second, we check that each sequent rule transforms the interpolation formula(s) of the premise sequent(s) into an interpolation formula of the conclusion sequent.

1. if $U, \alpha_1, \alpha_2 \xrightarrow{h} V$, then $U, \alpha \xrightarrow{h} V$.
2. if $U \xrightarrow{h} V, \beta_1, \beta_2$, then $U \xrightarrow{h} V, \beta$.
3. if $U \xrightarrow{h_1} V, \alpha_1$ and $U \xrightarrow{h_2} V, \alpha_2$, then $U \xrightarrow{h_1 \wedge h_2} V, \alpha$.
4. if $U, \beta_1 \xrightarrow{h_1} V$ and $U, \beta_2 \xrightarrow{h_2} V$, then $U, \beta \xrightarrow{h_1 \vee h_2} V$.
5. if $U, \neg\neg f \xrightarrow{h} V$, then $U, f \xrightarrow{h} V$.
6. if $U \xrightarrow{h} V, \neg\neg f$, then $U \xrightarrow{h} V, f$.
7. if $K^+(U, 1-c), \pi_0(c) \xrightarrow{h} \hat{K}^+(V, 1-c)$, then $U, \pi(c) \xrightarrow{(c)h} V$.
8. if $K(U, 1-c), \pi_0^+(c) \xrightarrow{h} \hat{K}(V, 1-c)$, then $U, \pi^+(c) \xrightarrow{(c)^+h} V$.
9. if $K^+(U, c) \xrightarrow{h} \hat{K}^+(V, c), \nu_0^+(c)$, then $U \xrightarrow{[c]^+h} V, \nu^+(c)$.
10. if $K(U, c) \xrightarrow{h} \hat{K}(V, c), \nu_0(c)$, then $U \xrightarrow{[c]h} V, \nu(c)$.
11. if $K^+(U, 0) \xrightarrow{h} \hat{K}^+(V, 0)$, then $U \xrightarrow{[0]^+h} V$ and $U \xrightarrow{(1)h} V$.

Next, we will consider the QML version of Beth's definability theorem which is proposed in (Beth 1953) for classical logic and generalized by Fitting (1983) for modal logic.

Definition 8. 1. Let p be a propositional variable and $f(p)$ be a wff containing p. We say there is an explicit definition of p from $f(p)$ in the QML if there is a formula g, in which p does not occur, such that $\models f(p) \supset (p \equiv g)$
2. $f(p)$ is said to define p implicitly in the QML if for each propositional variable q that does not occur in $f(p)$, $\models f(p) \wedge f(q) \supset (p \equiv q)$

Theorem 9 (Beth's Definability Theorem). *If $f(p)$ defines p implicitly in QML, then there is also an explicit definition of p from $f(p)$ in QML.*

Proof. By using Craig's interpolation lemma of QML, the proof is same as that for modal logics (Fitting 1983).

5 Conclusion

We have made two points in this chapter. First, we showed how to reformulate possibilistic reasoning as a kind of modal logic in order to enhance its expressive power. In the resultant language, QML, we can represent both high-order uncertainty and the interaction between weighted wffs and classical wffs. Secondly, we developed a Gentzen deduction method for QML that removes the restriction to clausal form. In summary, we have a sound and complete deduction system for general possibilistic reasoning for as long as the global assumption set is finite. Since the problem encountered in the real world is usually finite, the system reported in this paper should be practical. The only remaining problem is the efficiency of the method. It is generally believed that resolution is a more efficient deduction method in automated theorem proving. However, we believe that by using careful control strategies, Gentzen systems can also have good performance. The method developed by Wallen (1987) provides a good example.

References

Beth, E. W.: "On Padoa's method in the theory of definition," in: *Indag. Math.* 15 (1953) 330-339.

Craig, W.: "Linear reasoning. A new form of the Herbrand-Gentzen theorem," in: *Journal of Symbol Logic* 22 (1957a) 250-268.

Craig, W.: "Three uses of the Herbrand-Gentzen theorem in relating model theory to proof theory," in: *Journal of Symbol Logic* 22 (1957b) 269-285.

Dubois, D., J. Lang, and H. Prade: "Fuzzy sets in approximate reasoning.Part 2: Logical approache," in: *Fuzzy Sets and Systems*, 25th Anniversary volume, 1990.

Dubois, D., and H. Prade: "An introduction to possibilistic and fuzzy logics," in: P. Smets et al. (eds.), *Non-standard Logics for Automated Reasoning*, Academic Press (1988) 287-325.

Dubois, D., H. Prade, and C. Testmale: "In search of a modal system for possibility theory," in: *Proc. of ECAI-88* (1988) 501-506.

Farinas del Cerro, L., and A. Herzig: "A modal analysis of possibility theory," in: Kruse and Siegel (eds.), *Proc. of ECSQAU*. Springer Verlag, LNAI 548 (1991) 58-62.

Fitting, M.C.: *Proof Methods for Modal and Intuitionistic Logics*. Vol. 169 of Synthese Library, D. Reidel Publishing Company, 1983.

Gaifman, H.: "A theory of higher order probabilities," in: J.Y. Halpern (ed.), *Proceeding of 1st Conference on Theoretical Aspects of Reasoning about Knowledge*. Morgan Kaufmann, (1986) 275-292.

Gentzen, G.: "Investigation into logical deduction," in: M. E. Szabo (ed.), *The Collected Papers of Gerhard Gentzen* Amsterdam: North-Holland (1969) 68-131.

Liau, C. J., and I. P. Lin: "Quantitative modal logic and possibilistic reasoning," in: *Proc. of ECAI 92*, 1992.

Robinson, J.A.: "A machine oriented logic based on the resolution principle," in: *JACM* 12 (1965) 23-41.

Wallen, L.: "Matrix proofs for modal logics," in: *Proc. of IJCAI* (1987) 917-923.

Zadeh, L.A.: "Fuzzy sets as a basis for a theory of possibility," in: *Fuzzy Sets and Systems*, 1 (1978) 3-28.

A Model of Inductive Reasoning *

Peter A. Flach

Institute for Language Technology & Artificial Intelligence
Tilburg University, POBox 90153, 5000 LE Tilburg, Netherlands
E-mail: flach@kub.nl

Abstract. This paper presents a formal characterization of the process of inductive hypothesis formation. This is achieved by formulating minimal properties for inductive consequence relations. These properties are justified by the fact that they are sufficient to allow identification in the limit. By means of stronger sets of properties, we also define both standard and non-standard forms of inductive reasoning, and give an application of the latter.

1 Introduction

1.1 Motivation and scope

Induction is the process of drawing conclusions about all members of a certain set from knowledge about specific members of that set. For example, after observing a number of black crows, we might conclude inductively that all crows are black. Such a conclusion can never be drawn with absolute certainty, and an immediate question is: how is our confidence in it affected by observing the next black crow? This problem is known as the *justification problem* of induction, a problem with which philosophers of all times have wrestled without finding a satisfactory solution.

In this paper, we are concerned with a different but related problem: the formalisation of the process of *inductive hypothesis formation*. Which hypotheses are possible, given the available information? For instance, in the crows example the hypothesis 'all crows are black' is possible, but the hypothesis 'all crows are white' is not: it is refuted as soon as we observe one black crow. Moreover, once refuted, it will never become a possible hypothesis again, no matter how many crows are observed. The question is thus: what is the relation between sets of observations and possible hypotheses?

In order to address this question, we need a representation for observations and hypotheses. In our framework, both are represented by logical statements. Since the above definition of induction makes an explicit distinction between instances and sets of instances, a first-order logic seems most appropriate. However, it can be shown that many typical induction problems (like concept learning

---

* Part of this work has been carried out under ESPRIT III Basic Research Action 6020: *Inductive Logic Programming*.

from examples) remain essentially the same if we regard an instance as an equivalence class of indistinguishable objects. It follows that the distinction between instances and sets is not essential, at least not for typical induction problems. For instance, in many approaches to concept learning from examples, a so-called *attribute-value language* is used to describe both examples and concepts in terms of properties like colour, shape and size. Attribute-value languages have the expressive power of propositional logic.[2]

Since observations and hypotheses are logical statements, inductive hypothesis formation can be declaratively modeled as a consequence relation. We will study the properties of such inductive consequence relations, thereby applying techniques developed in other fields of non-standard logic, like non-monotonic reasoning (Gabbay 1985, Shoham 1987, Kraus et al. 1990), abduction (Zadrozny 1991) and belief revision (Gärdenfors 1988, Gärdenfors 1990). In the spirit of these works, we develop several systems of properties that inductive consequence relations might have. They model *different kinds* of induction, with the weakest system **I** giving *necessary conditions* for inductive consequence relations.

As a motivating example, let the observations be drawn from a database of facts about different persons, including their first and last names, and their parents first and last names. A typical inductive hypothesis would be 'every person's last name equals her father's last name'. Such a hypothesis, if adopted, would yield a procedure for finding a persons last name, given her father's last name. Now consider the statement 'every person has exactly one mother'. It is also an inductive hypothesis, but of a different kind. Specifically, it does not give a procedure for finding a person's mother (such a procedure does obviously not exist), but merely states her existence and uniqueness. While the first hypothesis can be seen as a *definition* of the last name of children, the second one is instead a *constraint* on possible models of the database. In the sections to follow, we will give practical examples of both forms of induction.

1.2 Terminology and notation

Suppose P is a computer program that performs inductive reasoning. That is, P takes a set of formulae α in some language L as input, and outputs inductive conclusions β. The main idea is to view P as constituting a consequence relation, i.e., a relation on $2^L \times L$, and to study the properties of this consequence relation. We will write $\alpha \mathrel{\mkern-5mu\not\mkern-9mu\sim} \beta$ whenever β is an inductive consequence of the premises α. A set of premises is often represented by a formula expressing their conjunction. The properties of $\mathrel{\mkern-5mu\not\mkern-9mu\sim}$ will be expressed by Gentzen-style inference rules in a meta-logic, following (Gabbay 1985, Kraus et al. 1990).

We will assume that L is a propositional language, closed under the logical connectives. Furthermore, we assume a set of models M for L, and the classical satisfaction relation $\models$ on $M \times L$. If $m \models \alpha$ for all $m \in M$, we write $\models \alpha$. We can

---

[2] For simplicity, the framework in this paper will be developed on the basis of propositional logic; we have no reason to believe, however, that its applicability cannot be extended to first-order logic.

implicitly introduce background knowledge by restricting M to a proper subset of all possible models. We will assume that $\models$ is compact, i.e., an infinite set of formulae is unsatisfiable if and only if every finite subset is.

In many practical cases premises and hypotheses are drawn from restricted sublanguages of L. Given a language L, an *inductive frame* is a triple $\langle \Gamma, \mathrel{\vert\!\sim}, \Sigma \rangle$, where $\Gamma \subseteq L$ is the set of possible observations, $\Sigma \subseteq L$ is the set of possible hypotheses, and $\mathrel{\vert\!\sim}$ is an inductive consequence relation on $2^L \times L$. We will assume that Γ is at least closed under conjunction.

The fact that in an inductive frame the consequence relation is defined on $2^L \times L$, rather than $2^\Gamma \times \Sigma$, reflects an important choice for a certain interpretation of $\mathrel{\vert\!\sim}$. Specifically, we chose to interpret $\alpha \mathrel{\vert\!\sim} \beta$ not just as 'β is an inductive consequence of α', but more generally as 'β is a possible hypothesis, given α'. In this way, our framework allows the study of not only inductive reasoning, but hypothetical reasoning in general.

As an example, take the property of Contraposition (if $\alpha \mathrel{\vert\!\sim} \beta$, then $\neg\beta \mathrel{\vert\!\sim} \neg\alpha$), which we will encounter later. This property can be understood as follows. Suppose I know that c is a crow (background knowledge), and the premise that c is black (α) allows the inductive hypothesis all crows are black (β), then the premiss that some crows are not black ($\neg\beta$) allows likewise the hypothesis that c is not black ($\neg\alpha$). It is quite possible that, in most inductive frames, this hypothesis is excluded from Σ[3]. This prohibits the interpretation of Contraposition as stating 'if $\alpha \mathrel{\vert\!\sim} \beta$ is an inductive argument, then so is $\neg\beta \mathrel{\vert\!\sim} \neg\alpha$'. Rather, it describes a property of hypothesis formation in general: $\neg\alpha$ is a possible hypothesis on the basis of $\neg\beta$, just like β is a possible hypothesis on the basis of α.

The plan of the paper is as follows. In Section 2, we define a minimal set of properties for inductive consequence relations, and we show that these properties are sufficient in the sense that they allow for a very general induction method. In Sections 3 and 4, we develop two, more or less complementary, kinds of inductive reasoning, and we give their main properties. We end the paper with some concluding remarks.

2 Identification in the Limit and I-relations

2.1 Identification in the limit

Inductive arguments are *defeasible*: an inductive conclusion might be invalidated by future observations. Thus, the validity of an inductive argument can only be guaranteed when complete information is available. A possible way to model complete information is by a sequence of formulae (possibly infinite), such that every incorrect hypothesis is eventually ruled out by a formula in the sequence. If an inductive reasoner reads in a finite initial segment of this sequence and outputs a correct hypothesis, it is said to have *finitely identified* the hypothesis. Since this is a fairly strong criterion, it is often weakened as follows: the inductive

---

[3] In fact, it looks more like an abductive hypothesis. We have argued elsewhere (Flach 1992) that there are close relations between abduction and induction.

reasoner is allowed to output as many hypotheses as wanted, but after finitely many guesses the hypothesis must be correct, and not abandoned afterwards. This is called *identification in the limit*. The difference with finite identification is, that the inductive reasoner does not know when the correct hypothesis has been attained. Details can be found in (Gold 1967).

We will redefine identification in the limit in terms of inductive consequence relations. Given a set of hypotheses Σ, the task is to identify an unknown $\beta \in \Sigma$ from a sequence of observations $\alpha_1, \alpha_2, \ldots$, such that $\{\alpha_1, \alpha_2, \ldots\} \mathrel{\vdash\mkern-11mu\sim} \beta$. The observations must be sufficient in the sense that they eventually rule out every non-intended hypothesis.

Definition 1 (Identification in the limit). Let $\langle \Gamma, \mathrel{\vdash\mkern-11mu\sim}, \Sigma \rangle$ be an inductive frame. Given a *target hypothesis* $\beta \in \Sigma$, a *presentation for* β is a (possibly infinite) sequence of observations $\alpha_1, \alpha_2, \ldots$ such that $\{\alpha_1, \alpha_2, \ldots\} \mathrel{\vdash\mkern-11mu\sim} \beta$. Given a presentation, an *identification algorithm* is an algorithm which reads in an observation α_j from the presentation and outputs a hypothesis β_j, for $j = 1, 2, \ldots$. The output sequence $\beta_1, \beta_2, \ldots$ is said to *converge* to β_n if for all $k \geq n$, $\beta_k = \beta_n$.

A presentation $\alpha_1, \alpha_2, \ldots$ for β is *sufficient* if for any hypothesis $\gamma \in \Sigma$ other than β it contains a *witness* α_i such that $\{\alpha_1, \alpha_2, \ldots, \alpha_i\} \mathrel{\not\vdash\mkern-11mu\sim} \gamma$. An identification algorithm is said to *identify* β *in the limit* if, given any sufficient presentation for β, the output sequence converges to β. An identification algorithm identifies Σ in the limit, if it is able to identify any $\beta \in \Sigma$ in the limit. □

Since we place induction in a logical context, it makes sense not to distinguish between logically equivalent hypotheses. That is, β is *logically* identified in the limit if the output sequence converges to β' such that $\models \beta' \leftrightarrow \beta$. A presentation for β is *logically* sufficient if it contains a witness for any γ such that $\not\models \gamma \leftrightarrow \beta$. In the sequel, we will only consider logical identification, and omit the adjective logical.

2.2 I-relations

After having defined identification in the limit in terms of inductive consequence relations, we now turn to the question: what does it take for inductive consequence relations to behave sensibly? We will first consider some useful properties, and then combine these properties into formal system **I**. We consider this system to be the weakest possible system defining inductive consequence relations.

The first two properties follow from the definition of identification in the limit. Suppose that α_i is a witness for γ, i.e., $\{\alpha_1, \alpha_2, \ldots, \alpha_i\} \mathrel{\not\vdash\mkern-11mu\sim} \gamma$, then any extended set of observations should still refute γ, i.e., $\{\alpha_1, \alpha_2, \ldots, \alpha_i\} \cup A \mathrel{\not\vdash\mkern-11mu\sim} \gamma$ for any $A \subseteq \Gamma$. Conversely, if $B \mathrel{\vdash\mkern-11mu\sim} \beta$ then also $B' \mathrel{\vdash\mkern-11mu\sim} \beta$ for any $B' \subseteq B$. Assuming that sets of observations can always be represented by their conjunction, this property can be stated as follows:

$$\frac{\models \alpha \rightarrow \beta, \; \alpha \mathrel{\vdash\mkern-11mu\sim} \gamma}{\beta \mathrel{\vdash\mkern-11mu\sim} \gamma} \tag{1}$$

Furthermore, observations cannot distinguish between logically equivalent hypotheses:

$$\frac{\models \beta \leftrightarrow \gamma, \alpha \mathbin{\vert\!\sim} \beta}{\alpha \mathbin{\vert\!\sim} \gamma} \tag{2}$$

The other two properties are not derived directly from identification in the limit. Instead, they describe the relation between observations and the hypotheses they confirm or refute. Here, the basic assumption is that induction aims to increase knowledge about some unknown intended model m_0. The observations are obtained from a reliable source, and are therefore true in m_0. On the other hand, hypotheses represent assumptions about the intended model. Together, observations and hypotheses can be used to make predictions about m_0. More specifically, suppose we have adopted hypothesis β on the basis of observations α, and let δ be a logical consequence of $\alpha \wedge \beta$, then we expect δ to be true in m_0. If the next observation conforms to our prediction, then we stick to β; if it contradicts our prediction, β should be refuted. These two principles can be expressed as follows.

$$\frac{\models \alpha \wedge \beta \rightarrow \delta, \alpha \mathbin{\vert\!\sim} \beta}{\alpha \wedge \delta \mathbin{\vert\!\sim} \beta} \tag{3}$$

$$\frac{\models \alpha \wedge \beta \rightarrow \delta, \alpha \mathbin{\vert\!\sim} \beta}{\alpha \wedge \neg\delta \mathbin{\vert\!\!\not\sim} \beta} \tag{4}$$

Note that the combination of these rules requires that $\alpha \wedge \beta$ is consistent: otherwise, we would have both $\models \alpha \wedge \beta \rightarrow \delta$ and $\models \alpha \wedge \beta \rightarrow \neg\delta$, and thus both $\alpha \wedge \delta \mathbin{\vert\!\sim} \beta$ (3) and $\alpha \wedge \delta \mathbin{\vert\!\!\not\sim} \beta$ (4). For technical reasons, the inconsistency of $\alpha \wedge \beta$ is not prohibited a priori. In the presence of the other rules, the application of rule (4) can be blocked in this case by adding the consistency of β as a premise (Theorem 4).

Rules (1) and (3) look pretty similar, and can probably be combined into a single rule. Rule (2) is clearly independent from the other rules. Rule (4) is not derivable from the other rules, but it may be if we add a weaker version. These considerations lead to the following system of rules.

Definition 2 (I-relations). The system **I** consists of the following four rules:

Conditional Reflexivity : $\dfrac{\not\models \neg\alpha}{\alpha \mathbin{\vert\!\sim} \alpha}$

Consistency : $\dfrac{\not\models \neg\alpha}{\neg\alpha \mathbin{\vert\!\!\not\sim} \alpha}$

Right Logical Equivalence : $\dfrac{\models \beta \leftrightarrow \gamma, \alpha \mathbin{\vert\!\sim} \beta}{\alpha \mathbin{\vert\!\sim} \gamma}$

Convergence : $\dfrac{\models \alpha \wedge \gamma \rightarrow \beta, \alpha \mathbin{\vert\!\sim} \gamma}{\beta \mathbin{\vert\!\sim} \gamma}$

If $\mathrel{\vdash\!\!\!\sim}$ is a consequence relation satisfying the rules of I, it is called an *I-relation*.
□

The following lemma gives two useful derived rules in this system.

Lemma 3. *The following rules are derived rules in system* **I***:*

$$\text{S}: \quad \frac{\not\models \neg\beta, \models \beta \rightarrow \alpha}{\alpha \mathrel{\vdash\!\!\!\sim} \beta}$$

$$\text{W}: \quad \frac{\not\models \neg\beta, \alpha \mathrel{\vdash\!\!\!\sim} \beta}{\not\models \beta \rightarrow \neg\alpha}$$

Proof. (S) Suppose $\not\models \neg\beta$ and $\models \beta \rightarrow \alpha$; by Conditional Reflexivity it follows that $\beta \mathrel{\vdash\!\!\!\sim} \beta$, and we conclude by Convergence.

(W) Suppose $\not\models \neg\beta$ and $\alpha \mathrel{\vdash\!\!\!\sim} \beta$; by Consistency it follows that $\neg\beta \mathrel{\not\!\!\vdash\!\!\!\sim} \beta$. By Convergence, it follows that $\not\models \alpha \wedge \beta \rightarrow \neg\beta$, i.e., $\not\models \beta \rightarrow \neg\alpha$. □

Rule S expresses that β is an inductive consequence of α if α is deductively entailed by β; rule W states that β shouldn't deductively entail $\neg\alpha$ if it is an inductive consequence of α (with β consistent in both cases). Rule S hints at the view of induction as reversed deduction, while rule W suggests a connection between induction and finding consistent extensions of a theory. By strengthening one of these rules, we obtain systems for one of two different kinds of induction: strong induction (Section 3) and weak induction (Section 4).

The following theorem shows that system **I** does what it was intended to do.

Theorem 4. *Rules (1)-(4) are derived rules in system* **I***.*

Proof. (1) Suppose $\models \alpha \rightarrow \beta$, i.e., $\models \alpha \wedge \gamma \rightarrow \beta$ and $\alpha \mathrel{\vdash\!\!\!\sim} \gamma$; we have $\alpha \mathrel{\vdash\!\!\!\sim} \beta$ by Convergence.

(2) Identical to Right Logical Equivalence.

(3) Suppose $\models \alpha \wedge \beta \rightarrow \delta$, i.e., $\models \alpha \wedge \beta \rightarrow \alpha \wedge \delta$ and $\alpha \mathrel{\vdash\!\!\!\sim} \beta$; by Convergence it follows that $\alpha \wedge \delta \mathrel{\vdash\!\!\!\sim} \beta$. Note that in the presence of rule (1), rule (3) is equivalent to Convergence, since the latter can also be derived from the former: suppose $\models \alpha \wedge \gamma \rightarrow \beta$ and $\alpha \mathrel{\vdash\!\!\!\sim} \gamma$, then by (3) $\alpha \wedge \beta \mathrel{\vdash\!\!\!\sim} \gamma$, and since $\models \alpha \wedge \beta \rightarrow \beta$, by (1) $\beta \mathrel{\vdash\!\!\!\sim} \gamma$. Since we already showed that (1) follows from Convergence, we conclude that Convergence exactly replaces (1) and (3).

(4) As said earlier, we prove this rule under the assumption that β is consistent. Suppose $\alpha \wedge \neg\delta \mathrel{\vdash\!\!\!\sim} \beta$, then by rule W $\not\models \beta \rightarrow \neg(\alpha \wedge \neg\delta)$, i.e., $\not\models \alpha \wedge \beta \rightarrow \delta$. From $\alpha \mathrel{\vdash\!\!\!\sim} \beta$ and the consistency of β, it follows that $\alpha \wedge \beta$ is consistent by rule W, as required. □

It should be noted that Conditional Reflexivity is nowhere used in the proof of Theorem 4. This indicates that it can be removed to obtain a truly minimal rule system for induction. However, system **I** possesses a nice symmetry, as shown by the next result.

Theorem 5. *Define* $\alpha \not\!\!\!\ll \beta$ *iff* $\neg\alpha \not\!\!\!\not\ll \beta$, *then* $\ll$ *is an I-relation iff* $\not\!\!\!\ll$ *is an I-relation.*

Proof. Using the rewrite rule $\alpha \ll \beta \Rightarrow \neg\alpha \not\!\!\!\ll \beta$, Conditional Reflexivity rewrites to Consistency and *vice versa*, while Right Logical Equivalence and Convergence rewrite to themselves. Since this rewrite rule is its own inverse, this proves the theorem in both directions. □

This duality will reappear later, as it provides the link between weak and strong induction (Section 4.2).

System **I** has been built on the basis of rules (1)–(4), which in turn were derived from the notion of identification in the limit. The following section rounds off this analysis by demonstrating how one could use inductive consequence relations for performing the perhaps most elementary form of identification: identification by enumeration.

2.3 Identification by enumeration

If we assume that the set of inductive hypotheses is countable, we can formulate a very simple and general identification algorithm (Algorithm 6). We enumerate all the possible hypotheses, and search this enumeration for a hypothesis that is an inductive consequence of the premises seen so far. We stick to this hypothesis until we encounter a new premise which, together with the previous premises, contradicts it: then we continue searching the enumeration.

Algorithm 6 (Identification by enumeration).
Input: a presentation $\alpha_1, \alpha_2, \ldots$ for a target hypothesis $\beta \in \Sigma$, and an enumeration $\beta_1, \beta_2, \ldots$ of all the formulae in Σ.
Output: a sequence of formulae in Σ.

>   begin
> $i := 1; k := 1;$
>   repeat
>   while $\{\alpha_j | j \leq i\} \not\!\!\!\ll \beta_k$ do $k := k + 1$;
>   output β_k;
> $i := i + 1$;
>   forever;
>   end.

Algorithm 6 is very powerful, but it has one serious drawback: the enumeration of hypotheses is completely unordered. Therefore, there is much duplication of work in checking hypotheses. There exist more practical versions of this algorithm, that can be applied if the set of hypotheses can be ordered. However, it is clear that if any search-based identification algorithm can achieve identification in the limit, identification by enumeration can as well, provided the inductive consequence relation is well-behaved. The following theorem states that I-relations are well-behaved in this sense.

Theorem 7. *Algorithm 6 performs identification in the limit if $\;\vert\!\!\!\sim\;$ is an I-relation.*

Proof. Let α denote the entire presentation, and let β be the target hypothesis, i.e., $\alpha \mathrel{\vert\!\!\!\sim} \beta$. Furthermore, let β_n be the first formula in the enumeration, such that $\models \beta_n \leftrightarrow \beta$. We will show that the output sequence converges to β_n if α is sufficient for β.

Suppose β_k, $k < n$ precedes β_n in the presentation. By assumption, $\not\models \beta_k \leftrightarrow \beta$; if the presentation is sufficient, there will be a witness α_i such that $\{\alpha_1, \alpha_2, \ldots, \alpha_i\} \mathrel{\not\vert\!\!\!\sim} \beta_k$, so β_k will be discarded.

Since $\models \beta_n \leftrightarrow \beta$ and $\alpha \mathrel{\vert\!\!\!\sim} \beta$, it follows by Right Logical Equivalence that $\alpha \mathrel{\vert\!\!\!\sim} \beta_n$. By Convergence, $\alpha' \mathrel{\vert\!\!\!\sim} \beta_n$ for every initial segment α'. Therefore β_n is never discarded. $\qquad\square$

Note that this proof only mentions the rules Right Logical Equivalence and Convergence. As said before, Consistency is needed to ensure that the presentation and the hypothesis can be combined in a meaningful way; Conditional Reflexivity is only needed to preserve the symmetry expressed in Theorem 5.

Induction would be infeasible if it could only be achieved by enumerating hypotheses. Likewise, inductive consequence relations would be useless if nothing stronger than I-relations would exist. In the following two sections, we present two families of inductive consequence relations, the first based on a view of induction as reversed deduction, and the second based on a view of induction as trying to extend the observations consistently. We will thereby adopt a terminology that is more familiar in the field of Machine Learning: observations are called *examples*, and inductive hypotheses are also called *explanations* of the examples.

3 Strong Induction

A *strong* inductive consequence relation is an I-relation that satisfies the following rule:

$$\mathbf{S'}: \quad \frac{\models \beta \rightarrow \alpha}{\alpha \mathrel{\vert\!\!\!\sim} \beta}$$

This is a strengthening of the derived rule S in system I (Lemma 3). The intuition behind strong induction is, that it is a form of reversed deduction in some underlying logic or *base logic*. Rule S′ states, that this base logic should allow all valid classical deductions. The base logic might also allow deductions that are classically invalid, but (for instance) plausible. Rule S′ is a strengthening of S, because it does not require β to be consistent. In general, inconsistent inductive hypotheses are not very interesting; they arise as a borderline case, similar to tautologies that are deductive consequences of any set of premises. In the present context, this borderline case is instrumental in distinguishing strong induction from weak induction, as we will see in Section 4.

Rule S' can be derived if we strengthen Conditional Reflexivity to

Reflexivity : $\alpha \mathrel{\vcenter{\hbox{$\vert\!\!\sim$}}} \alpha$

Thus, an I-relation is a strong inductive consequence relation iff it satisfies Reflexivity. In the two rule systems for strong induction to come, Reflexivity replaces Conditional Reflexivity.

3.1 The system SC

The weakest system of rules for strong induction is called **SC**, which stands for Strong induction with a Cumulative base logic. For inductive consequence relations, cumulativity means that if $\beta \mathrel{\vcenter{\hbox{$\vert\!\!\sim$}}} \gamma$, the hypotheses γ and $\beta \wedge \gamma$ inductively explain exactly the same facts. This principle can be expressed by two rules: Right Cut and Right Extension.

Definition 8 (The system **SC**). The system **SC** consists of the following six rules:

Reflexivity : $\alpha \mathrel{\vcenter{\hbox{$\vert\!\!\sim$}}} \alpha$

Consistency : $\dfrac{\not\models \neg \alpha}{\neg \alpha \mathrel{\vcenter{\hbox{$\vert\!\!\not\sim$}}} \alpha}$

Right Logical Equivalence : $\dfrac{\models \beta \leftrightarrow \gamma,\ \alpha \mathrel{\vcenter{\hbox{$\vert\!\!\sim$}}} \beta}{\alpha \mathrel{\vcenter{\hbox{$\vert\!\!\sim$}}} \gamma}$

Convergence : $\dfrac{\models \alpha \wedge \gamma \rightarrow \beta,\ \alpha \mathrel{\vcenter{\hbox{$\vert\!\!\sim$}}} \gamma}{\beta \mathrel{\vcenter{\hbox{$\vert\!\!\sim$}}} \gamma}$

Right Cut : $\dfrac{\alpha \mathrel{\vcenter{\hbox{$\vert\!\!\sim$}}} \beta \wedge \gamma,\ \beta \mathrel{\vcenter{\hbox{$\vert\!\!\sim$}}} \gamma}{\alpha \mathrel{\vcenter{\hbox{$\vert\!\!\sim$}}} \gamma}$

Right Extension : $\dfrac{\alpha \mathrel{\vcenter{\hbox{$\vert\!\!\sim$}}} \gamma,\ \beta \mathrel{\vcenter{\hbox{$\vert\!\!\sim$}}} \gamma}{\alpha \mathrel{\vcenter{\hbox{$\vert\!\!\sim$}}} \beta \wedge \gamma}$

□

Right Cut expresses that a part of an inductive hypothesis, which inductively explains another part, may be cut away from the hypothesis. In the presence of rule S', it is a strengthening of Convergence. Right Extension states that an inductive hypothesis may be extended by some of the things it explains.

These latter two rules may look suspicious, because β takes the role of both example and hypothesis. For instance, Right Extension might not be applicable in a particular inductive frame, because $\beta \notin \Sigma$. The reader will recall the discussion in Section 1.2, where it was argued that even if this is so, such rules may describe useful properties of the process of (inductive) hypothesis formation. Here we encounter a case in point, because the two new rules interact to produce a rule that is satisfied in any inductive frame of which the consequence relation satisfies the rules of **SC**.

Lemma 9. *In* SC, *the following rule can be derived:*

$$\text{Compositionality}: \quad \frac{\alpha \mathrel{|\!\sim} \gamma,\; \beta \mathrel{|\!\sim} \gamma}{\alpha \wedge \beta \mathrel{|\!\sim} \gamma}$$

Proof. Suppose $\alpha \mathrel{|\!\sim} \gamma$ and $\beta \mathrel{|\!\sim} \gamma$; by Right Extension we have $\alpha \mathrel{|\!\sim} \beta \wedge \gamma$. Also, because $\alpha \wedge \beta \wedge \gamma \models \alpha \wedge \beta$, we have $\alpha \wedge \beta \mathrel{|\!\sim} \alpha \wedge \beta \wedge \gamma$ by rule S'. Using Right Cut gives $\alpha \wedge \beta \mathrel{|\!\sim} \beta \wedge \gamma$, and since by assumption $\beta \mathrel{|\!\sim} \gamma$, we can cut away β from the righthand side to get $\alpha \wedge \beta \mathrel{|\!\sim} \gamma$. □

Compositionality states that if an inductive hypothesis explains two examples separately, it also explains them jointly. It can be employed to speed up enumerative identification algorithms. Recall that in Algorithm 6 a new hypothesis must be checked against the complete set of previously seen examples. If we already know that the new hypothesis inductively explains some subset of those examples, then by Compositionality the remaining examples can be tested in isolation.

Furthermore, if the search strategy guarantees that the new hypothesis explains all the examples explained by the previous hypothesis, then we only need to test it against the last example which refuted the previous hypothesis. This requires an ordering of the hypothesis space, which in turn requires monotonicity of the base logic. This results in the following stronger system.

3.2 The system SM

There are several ways to define monotonicity of the base logic, for instance by adopting transitivity or contraposition. The next system is obtained by extending SC with a rule for contraposition (this makes transitivity a derived property).

Definition 10 (The system SM). The system SM consists of the rules of SC plus the following rule:

$$\text{Contraposition}: \quad \frac{\alpha \mathrel{|\!\sim} \beta}{\neg \beta \mathrel{|\!\sim} \neg \alpha}$$

□

As the following lemma shows, this results in a considerably more powerful system.

Lemma 11. *In* SM, *the following rules can be derived:*

$$\text{Explanation Strengthening}: \quad \frac{\models \gamma \rightarrow \beta,\; \alpha \mathrel{|\!\sim} \beta}{\alpha \mathrel{|\!\sim} \gamma}$$

$$\text{Explanation Updating}: \quad \frac{\models \gamma' \rightarrow \gamma,\; \alpha \mathrel{|\!\sim} \gamma,\; \beta \mathrel{|\!\sim} \gamma'}{\alpha \wedge \beta \mathrel{|\!\sim} \gamma'}$$

Proof. (Explanation Strengthening) Suppose $\models \gamma \to \beta$ and $\alpha \mathrel{|\kern-0.4em\sim} \beta$; by Contraposition, it follows that $\neg\beta \mathrel{|\kern-0.4em\sim} \neg\alpha$. Convergence gives $\neg\gamma \mathrel{|\kern-0.4em\sim} \neg\alpha$, which finally results in $\alpha \mathrel{|\kern-0.4em\sim} \gamma$ by Contraposition.

(Explanation Updating) Suppose $\models \gamma' \to \gamma$ and $\alpha \mathrel{|\kern-0.4em\sim} \gamma$; by Explanation Strengthening we have $\alpha \mathrel{|\kern-0.4em\sim} \gamma'$. Assuming $\beta \mathrel{|\kern-0.4em\sim} \gamma'$, this gives $\alpha \wedge \beta \mathrel{|\kern-0.4em\sim} \gamma'$ by Compositionality. □

Explanation Strengthening expresses that any γ logically implying some inductive explanation β of a set of examples α is also an explanation of α. Consequently, the set of inductive explanations of a given set of examples is completely determined by its weakest elements according to logical implication. Since logical implication is reflexive and transitive, it is a quasi-ordering on Σ, which can be turned into a partial ordering by considering equivalence classes of logically equivalent formulae (in other words, the Lindenbaum algebra of Σ).

Explanation Updating is a combination of Explanation Strengthening and Compositionality, which shows how to employ this ordering in identification algorithms. It states that if γ is a hypothesis explaining the examples α seen so far but not the next example β, it can be replaced by some γ' which (*i*) logically implies γ and (*ii*) explains β. This clearly shows that we don't need to test the new hypothesis γ' against the previous examples α.

The properties expressed by these rules have been used in many AI-approaches to inductive reasoning (Mitchell 1982, Shapiro 1983). The results in this section have been presented to show how they can be derived systematically within our framework. For instance, we have shown that an important property like Explanation Strengthening requires monotonicity of the base logic.

4 Weak Induction

The ideas described in this section have been the main motivating force for the research reported in this paper. While induction and deduction are closely related, they can be related in more than one way. Weak induction provides an alternative for strong induction, which only considers inductive hypotheses from which the examples are provable. Weak induction aims at supplementing the examples with knowledge which is only implicitly contained in those examples.

In Section 1.1, we provided an example of weak induction: inferring 'every person has exactly one mother' from a collection of facts. This is not a strong inductive argument, since the induced rule does not entail the facts (regardless of the base logic). Rather, the rule does not contradict the facts: it should not imply their falsity. That is, weak inductive consequence relations satisfy the following rule:

$$\mathbf{W'} : \quad \frac{\alpha \mathrel{|\kern-0.4em\sim} \beta}{\not\models \beta \to \neg\alpha}$$

Note that this disallows the possibility that β is inconsistent, showing that some strong inductive consequence relations are not weak inductive consequence relations.

Rule W' is a strengthening of rule W, and can be derived in I if we strengthen Consistency to

Weak Reflexivity : $\neg\alpha \not\!\!\!\sim \alpha$

Weak Reflexivity expresses that an inductive hypothesis never explains its negation. It replaces Consistency in the two systems for weak induction we will consider in this section.

4.1   The system WC

The weakest system for weak inductive reasoning is called **WC**. It models weak induction with a cumulative base logic. The principle of cumulativity for weak inductive consequence relations is stated as follows: if $\neg\beta \not\!\!\!\sim \gamma$, then β can be added to the inductive hypothesis γ without changing the set of examples it explains. This principle requires weak counterparts of the corresponding rules in **SC**.

Definition 12 (The system **WC**). The system **WC** consists of the following six rules:

Conditional Reflexivity : $\dfrac{\not\models \neg\alpha}{\alpha \mathrel{\vrule height 1.6ex depth 0pt width 0pt}\!\!\sim \alpha}$

Weak Reflexivity : $\neg\alpha \not\!\!\!\sim \alpha$

Right Logical Equivalence : $\dfrac{\models \beta \leftrightarrow \gamma, \alpha \mathrel{\vrule height 1.6ex depth 0pt width 0pt}\!\!\sim \beta}{\alpha \mathrel{\vrule height 1.6ex depth 0pt width 0pt}\!\!\sim \gamma}$

Convergence : $\dfrac{\models \alpha \wedge \gamma \rightarrow \beta, \alpha \mathrel{\vrule height 1.6ex depth 0pt width 0pt}\!\!\sim \gamma}{\beta \mathrel{\vrule height 1.6ex depth 0pt width 0pt}\!\!\sim \gamma}$

Weak Right Cut : $\dfrac{\alpha \mathrel{\vrule height 1.6ex depth 0pt width 0pt}\!\!\sim \beta \wedge \gamma, \neg\beta \not\!\!\!\sim \gamma}{\alpha \mathrel{\vrule height 1.6ex depth 0pt width 0pt}\!\!\sim \gamma}$

Weak Right Extension : $\dfrac{\alpha \mathrel{\vrule height 1.6ex depth 0pt width 0pt}\!\!\sim \gamma, \neg\beta \not\!\!\!\sim \gamma}{\alpha \mathrel{\vrule height 1.6ex depth 0pt width 0pt}\!\!\sim \beta \wedge \gamma}$

$\square$

In this system, we could derive a weak counterpart to Compositionality, expressing that an example, of which the negation is not explained, can be added to the premises. However, this does not express a very useful property. In general, Compositionality itself does not apply to weak inductive reasoning. Consequently, we must always store all previously seen examples, and check them each time we switch to a new inductive hypothesis (an illustration of this will be provided in Section 4.3).

4.2 The system WM

In a monotonic base logic, β does not entail $\neg\alpha$ if and only if α does not entail $\neg\beta$. This property means that a weak inductive consequence relation based on a monotonic base logic is *symmetric*. Again, we stress that although Symmetry is obviously not a property of any form of inductive reasoning, it may be a useful property of the inductive consequence relation involved in weak inductive reasoning.

Definition 13 (The system **WM**). The system **WM** consists of the rules of **WC** plus the following rule:

$$\textbf{Symmetry}: \quad \frac{\alpha \mathrel{\rlap{\,/}{\vdash}} \beta}{\beta \mathrel{\rlap{\,/}{\vdash}} \alpha}$$

$\Box$

Similar to **SM**, **WM** induces an ordering of the hypothesis space that can be exploited in enumerative identification algorithms. Search will however proceed in the opposite direction of logically weaker formulae.

Lemma 14. *In* **WM**, *the following rule can be derived:*

$$\textbf{Explanation Weakening}: \quad \frac{\models \beta \to \gamma, \alpha \mathrel{\rlap{\,/}{\vdash}} \beta}{\alpha \mathrel{\rlap{\,/}{\vdash}} \gamma}$$

Proof. Suppose $\models \beta \to \gamma$ and $\alpha \mathrel{\rlap{\,/}{\vdash}} \beta$; by Symmetry, it follows that $\beta \mathrel{\rlap{\,/}{\vdash}} \alpha$. Convergence gives $\gamma \mathrel{\rlap{\,/}{\vdash}} \alpha$, which finally results in $\alpha \mathrel{\rlap{\,/}{\vdash}} \gamma$ by Symmetry. $\Box$

In fact, **SM** and **WM** are interdefinable in the following sense.

Lemma 15. *Define* $\alpha \mathrel{\rlap{\,/\!/}{\vdash}} \beta$ *iff* $\neg\alpha \mathrel{\rlap{\,/\!/}{\not\vdash}} \beta$, *then* $\mathrel{\rlap{\,/}{\vdash}}$ *satisfies the rules of* **SM** *iff* $\mathrel{\rlap{\,/\!/}{\vdash}}$ *satisfies the rules of* **WM**.

Proof. Using the rewrite rule $\alpha \mathrel{\rlap{\,/}{\vdash}} \beta \Rightarrow \neg\alpha \mathrel{\rlap{\,/\!/}{\not\vdash}} \beta$, each rule of **SM** rewrites (after re-arranging) to a rule of **WM**: Reflexivity rewrites to Weak Reflexivity, Consistency rewrites to Conditional Reflexivity, Convergence and Right Logical Equivalence rewrite to themselves, Right Cut to Weak Right Extension, Right Extension to Weak Right Cut, and Contraposition rewrites to Symmetry. $\Box$

We encountered this transformation before, when we noted that it leaves system **I** invariant (Theorem 5).

4.3 An application of weak induction

In this section, we will illustrate the usefulness of weak induction by applying it to the problem of inducing integrity constraints in a deductive database. Tuples of a database relation (i.e., ground facts) play the role of examples, and hypotheses are integrity constraints on this relation. The induction algorithm is fully described in (Flach 1990), and has also been implemented. In (Flach 1993), we describe how the induced integrity constraints can be utilised for restructuring the database in a more meaningful way. In the current implementation, hypotheses are restricted to functional and multivalued dependencies between attributes.

Suppose `child` is a relation with five attributes: child's first name, father's first and last name, and mother's first and last name. Given the tuples listed in Table 1, we might for example be interested in the attributes that functionally determine the mother's last name (a socalled *functional dependency*).

Table 1. A database relation

```
child(john,frank,johnson,mary,peterson).
child(peter,frank,johnson,mary,peterson).
child(john,robert,miller,gwen,mcintyre).
child(ann,john,miller,dolly,parton).
child(millie,frank,miller,dolly,mcintyre).
```

Two such dependencies that are satisfied in Table 1 are:[4]

$$\texttt{child(N, \_, FL, \_, ML1)} \land \texttt{child(N, \_, FL, \_, ML2)} \rightarrow \texttt{ML1} = \texttt{ML2}$$

$$\texttt{child(\_, FF, FL, \_, ML1)} \land \texttt{child(\_, FF, FL, \_, ML2)} \rightarrow \texttt{ML1} = \texttt{ML2}$$

The first formula states that child's first name and father's last name determine mother's last name, and the second formula says that father's first and last names determine mother's last name. Note that these formulae are not logical consequences of the tuples, nor are the tuples logical consequences of the formulae.

How would we induce these dependencies? According to Explanation Weakening, we can start with the strongest hypothesis: all mothers have the same last name (it is determined by the empty set of attributes). This is expressed by the following formula:

$$\texttt{child(\_, \_, \_, \_, ML1)} \land \texttt{child(\_, \_, \_, \_, ML2)} \rightarrow \texttt{ML1} = \texttt{ML2}$$

Since this formula is inconsistent with the tuples in Table 1, we will make minimal changes in order to get weaker constraints, which is done by unifying variables on the lefthand side.

For instance, the first and third tuple lead to the following false formula:

---

[4] We follow the Prolog conventions: all variables are universally quantified, and the underscores denote unique variables.

```
child(john, frank, johnson, mary, peterson) ∧
child(john, robert, miller, gwen, mcintyre)
    → peterson = mcintyre
```

The formula is false because = is interpreted as syntactical identity. It shows how we can make minimal changes to the original formula: by unifying variables in those positions for which the tuples have different values. This leads to the following three hypotheses:

$$child(\_, FF, \_, \_, ML1) \land child(\_, FF, \_, \_, ML2) \rightarrow ML1 = ML2$$

$$child(\_, \_, FL, \_, ML1) \land child(\_, \_, FL, \_, ML2) \rightarrow ML1 = ML2$$

$$child(\_, \_, \_, MF, ML1) \land child(\_, \_, \_, MF, ML2) \rightarrow ML1 = ML2$$

Each of these hypotheses is again tested for consistency with the data.

If we search in a breadth-first fashion, we will eventually encounter all sets of attributes that determine the mother's last name. Note that, any time we switch to a new hypothesis, we have to check it against the *complete* set of tuples (Compositionality does not hold).

In this setting, there are rather strong restrictions on both Γ (the tuples) and Σ (the functional dependencies). They are needed to ensure convergence of the induction process, and also block properties like Symmetry. On the other hand, Γ should be rich enough to allow sufficient presentations for any hypothesis in Σ (they should form what Shapiro (1983) calls an *admissible pair*). For instance, let Σ be the set of *multivalued dependencies* that hold for a given database relation. An example of such a dependency is

```
child(N1, FF1, FL1, MF, ML) ∧
child(N2, FF2, FL2, MF, ML)
    → child(N1, FF2, FL2, MF, ML)
```

which states that children have all the fathers of any child of a certain mother. Such dependencies can be learned in exactly the same way as functional dependencies. The point is, that Γ should now contain positive *and negative* ground facts, since a given multivalued dependency can only be refuted by two tuples in the relation and one tuple known to be not in the relation.

5 Conclusion and Future Work

The contributions presented in this paper are twofold. First of all, we have given minimal conditions for inductive consequence relations, which are powerful enough to allow identification in the limit. On the other hand, these conditions are liberal enough as to leave room for non-standard forms of inductive reasoning. Our second contribution lies in identifying weak induction as such a non-standard form of induction. We have illustrated the usefulness of weak induction by applying it to the problem of inducing integrity constraints in a deductive database.

We are currently working on the model-theoretic counterpart of the proof-theoretic characterisation presented in this paper. On the basis of such a model theory, we should be able to substantiate our claim that the rules of system I indeed express necessary properties of an inductive consequence relation. Another topic of interest is the study of induction with respect to other base logics, such as modal, temporal and intuitionistic logics.

Acknowledgements

I would like to thank John-Jules Meyer, Yao-Hua Tan, Cees Witteveen, and the two anonymous referees for their insightful remarks.

References

Flach, P.A.: "Inductive characterisation of database relations," in: Z.W. Ras, M. Zemankowa & M.L. Emrich (eds.), *Proc. International Symposium on Methodologies for Intelligent Systems*, Amsterdam: North-Holland (1990) 371-378. Full version appeared as ITK Research Report no. 23.

Flach, P.A.: "An analysis of various forms of 'jumping to conclusions'," in: K.P. Jantke (ed.), *Analogical and Inductive Inference AII'92*, Lecture Notes in Artifical Intelligence 642, Berlin: Springer Verlag (1992) 170-186.

Flach, P.A.: "Predicate invention in Inductive Data Engineering," in: P.B. Bradzil (ed.), *Proc. European Conference on Machine Learning ECML'93*, Lecture Notes in Artifical Intelligence 667, Berlin: Springer Verlag (1993) 83-94.

Gabbay, D.M.: "Theoretical foundations for non-monotonic reasoning in expert systems," in: K.R. Apt (ed.), *Logics and Models of Concurrent Systems*, Berlin: Springer Verlag (1985) 439-457.

Gärdenfors. P.: *Knowledge in Flux*, Cambridge, MA: The MIT Press, 1988.

Gärdenfors, P.: "Belief revision and nonmonotonic logic: two sides of the same coin?" in: *Proc. Ninth European Conference on AI*, London: Pitman (1990) 768-773.

Gold, E.M.: "Language identification in the limit," in: *Information and Control* 10 (1967), 447-474.

Kraus, S., D. Lehmann and M. Magidor: "Nonmonotonic reasoning, preferential models and cumulative logics," in: *Artificial Intelligence* 44 (1990) 167-207.

Mitchell, T.M.: "Generalization as search," in:*Artificial Intelligence* 18 (1982) 2, 203-226.

Shapiro, E.Y.: *Algorithmic program debugging*. Cambridge MA: MIT Press, 1983.

Shoham, Y.: "A semantical approach to nonmonotonic logics," in: *Proc. Eleventh International Joint Conference on AI*, Los Altos, CA: Morgan Kaufmann (1987) 1304-1310.

Zadrozny, W.: *On rules of abduction*. IBM Research Report (August 1991).

Automated Reasoning with Uncertainties

Flávio S. Corrêa da Silva[1], Dave S. Robertson[2], Jane Hesketh[2]

[1] Instituto de Matemática e Estatística – Cid. Universitária "ASO" – PO Box 20570
– 01498 São Paulo SP Brazil – E-mail: fcs@ime.usp.br
[2] Dept. of Artificial Intelligence – Univ. of Edinburgh – 80 South Bridge – Edinburgh
– Scotland EH1 1HN – E-mail: {dr/jane}@aisb.ed.ac.uk

Abstract. In this work we assume that uncertainty is a multifaceted concept and present a system for automated reasoning with multiple representations of uncertainty.

We present a case study on developing a computational language for reasoning with uncertainty, starting with a semantically sound and computationally tractable language and gradually extending it with specialised syntactic constructs to represent measures of uncertainty, while preserving its unambiguous semantic characterization and computability properties. Our initial language is the language of normal clauses with SLDNF as the inference rule, and we select three specific facets of uncertainty for our study: *vagueness, statistics* and *degrees of belief.*

The resulting language is *semantically sound* and *computationally tractable.* It also admits relatively efficient implementations employing $\alpha - \beta$ *pruning* and *caching.*

1 Introduction

When reasoning we frequently use uncertain information, i.e., information that is incomplete, vague, only partially reliable or based on statistical associations. Hence, when building automated reasoning systems, we frequently need tools and mechanisms to represent uncertainty.

In this work we present a *system for automated reasoning with multiple representations of uncertainty:* uncertainty is a multifaceted concept, and because of this there are several techniques for measuring it. Our focus in this work is on problems which present more than one of these facets, problems, hence, for which it is important to differentiate several kinds of uncertainty.

We present a case study on developing a computational language for reasoning with uncertainty, starting with a semantically sound and computationally tractable language and gradually extending it to represent measures of uncertainty, preserving its unambiguous semantic characterisation and computability properties. Our initial language is the language of *normal clauses with SLDNF as the inference rule* (i.e., the language of pure PROLOG (Kunen 1989)), which is expressive enough to represent a significant portion of first-order logic, admits computationally tractable implementations, and has a well defined formal semantics.

We select three specific facets of uncertainty for our study, which are not exhaustive but cover many situations found in practical problems. These facets are (i) *vagueness*, which describes the extent to which a non-categorical statement is true - a vague predicate is one which truth-value admits intermediate values between *true* and *false* (e.g., the predicate "fat" qualifying the weight of a person); *statistics*, which describes the likelihood of selecting an element or class of elements belonging to the domain of discourse; and *degrees of belief*, which describe the belief apportioned to statements represented by sentences in our language.

With each of these facets we associate a specific measure. We associate *fuzzy measures* to vagueness (Dubois and Prade 1988), *probabilities on the domain* to statistics (Bacchus 1990b) and *probabilities on possible worlds* to degrees of belief (Corrêa da Silva and Bundy 1991, Nilsson 1986, Shafer 1976). The resulting language is *semantically sound* and *computationally tractable*, using in its implementation the standard optimisation techniques of $\alpha - \beta$ *pruning* and *caching*.

In Section 2 we review the main concepts of fuzzy set theory and probability theory which are used throughout the rest of the work. In Section 3 we introduce a logic programming language that can treat fuzzy predicates. The language treats negation by finite failure and is sound with respect to completed models. In Section 4 we extend this language to deal with probabilities on the domain. The language implements a significant subset of the logic L_p (Bacchus 1990b), extended with fuzzy predicates. The logic L_p was known to have computable subsets, but we are not aware of any previous implementations of it. In Section 5 we introduce the concepts of possible worlds and degrees of belief to the language. These concepts are introduced in a way that establishes close relations between our formalism and other well-known formalisms like Incidence Calculus (Bundy 1985, Corrêa da Silva and Bundy 1991), Probabilistic Logic (Nilsson 1986) and the Dempster-Shafer Theory of Evidence (Shafer 1976, Fagin and Halpern 1989a). Finally, Section 7 summarises and concludes this work.

2 General Definitions

In this section we introduce the concept of *fuzzy sets and relations*, to be used later in the interpretation of fuzzy sentences. Then we review the basic concepts of *probability theory* and its extensions to fuzzy events.

2.1 Fuzzy measures

A fuzzy membership function measures the degree to which an element belongs to a subset or, alternatively, the degree of similarity between the class (subset) to which an element belongs and a reference class. Formally, a fuzzy subset F of a referential set D is defined by an arbitrary mapping $\mu_F : D \rightarrow [0,1]$, in which, for an element $d \in D$, $\mu_F(d) = 1$ corresponds to the intuitive notion that $d \in F$ and $\mu_F(d) = 0$ to the notion that $d \notin F$ (Dubois and Prade 1989).

Set-theoretic operations can be extended to fuzzy sets by means of *triangular norms and conorms*. A *triangular norm* is any function $T : [0,1] \times [0,1] \rightarrow [0,1]$ such that:

- $T(x, 1) = x$ (boundary condition);
- $x_1 \leq x_2, y_1 \leq y_2 \Rightarrow T(x_1, y_1) \leq T(x_2, y_2)$ (monotonicity);
- $T(x, y) = T(y, x)$ (commutativity);
- $T(T(x, y), z) = T(x, T(y, z))$ (associativity).

The *conorm of a triangular norm* is the function $S : [0, 1] \times [0, 1] \rightarrow [0, 1]$ defined by:

$$S(x, y) = 1 - T(1 - x, 1 - y).$$

Furthermore, following (Dubois and Prade 1989), any function $C : [0, 1] \rightarrow [0, 1] : C(\mu_F(d)) = 1 - \mu_F(d)$ obeys the requirements as extension of complementation.

Not all algebraic properties of set operations are necessarily shared by triangular norms and conorms. In fact, as presented in (Klement 1982), the only norms and conorms that are also *distributive* and *idempotent* - i.e., that obey the following rules:

$$-\left\{ \begin{array}{l} S(x, T(y, z)) = T(S(x, y), S(x, z)) \\ T(x, S(y, z)) = S(T(x, y), T(x, z)) \end{array} \right\} \text{ (distributivity)};$$
- $T(x, x) = x$ and $S(x, x) = x$ (idempotency)

- are $T = min$ and $S = max$ - known as Zadeh's triangular norms and conorms. Henceforth, in order to keep fuzzy set operations as close as possible to conventional set operations, we adopt the following functions as our extended set operations of intersection, union and complementation:
- intersection: $\mu_{A \cap B}(x) = min\{\mu_A(x), \mu_B(x)\}$;
- union: $\mu_{A \cup B}(x) = max\{\mu_A(x), \mu_B(x)\}$;
- complementation: $\mu_{\neg A}(x) = 1 - \mu_A(x)$.
These are the most commonly used definitions of fuzzy set operations.

2.2 Probability measures

Given a finite set D, an *algebra* χ_D on D is a set of subsets of D such that (i) $D \in \chi_D$; (ii) $A \in \chi_D \Rightarrow \neg A \in \chi_D$; (iii) $A, B \in \chi_D \Rightarrow A \cup B \in \chi_D$.

Any subset of D is called an *event* on D. Events belonging to χ_D are called *measurable events*.

The *basis* χ'_D of an algebra χ_D is the subset of χ_D such that (i) $\{\} \notin \chi'_D$; (ii) $A, B \in \chi'_D \Rightarrow A \cap B = \{\}$; (iii) $K \in \chi_D \Rightarrow \exists A_1, ..., A_n \in \chi'_D : K = \bigcup_1^n A_i$.

A *probability measure* on χ_D is a function $\mathcal{P} : \chi_D \rightarrow [0, 1]$ such that (i) $\mathcal{P}(D) = 1$ (total probability); (ii) $A \cap B = \{\} \Rightarrow \mathcal{P}(A \cup B) = \mathcal{P}(A) + \mathcal{P}(B)$ (finite additivity).

As shown in (Fagin and Halpern 1989a), once $\mathcal{P}$ is defined for χ'_D it can be extended to the whole algebra by finite additivity. This is useful as we can specify a probability measure by defining its value only for the elements of χ'_D.

Given two measurable events $A, B \in \chi_D$, the *conditional probability* $\mathcal{P}(A|B)$ is defined as:

$$P(A|B) = \begin{cases} \frac{P(A \cap B)}{P(B)}, & P(B) \neq 0 \\ 0, & P(B) = 0 \end{cases}$$

Two measurable events A, B are called *independent* iff $P(A|B) = P(A)$ which, as a corollary, gives that $P(A \cap B) = P(A) \times P(B)$.

The set D can be partitioned into m subsets $D_1, ..., D_m$ such that (i) $D_i \cap D_j = \{\}, i, j = 1, ..., m$, and (ii) $D = \bigcup_1^m D_i$.

We can have independent algebras (two algebras χ_1 and χ_2 are *independent* iff each event $X_i \in \chi_1$ is independent of every event $Y_j \in \chi_2$ and vice-versa) χ_{Di} and probability measures P_i for each set D_i. If we assume that all events in each χ_{Di} are *pairwise independent*, we can extend measures to cartesian products of the sets D_i of a partition of D: the cartesian product of a collection of bases of algebras of elements $D_1, ..., D_m$ of a partition of D is the basis χ' of an algebra of the cartesian product of the sets $D_1, ..., D_m$, and the measure P on the corresponding algebra χ is defined as:

- $P : \chi \rightarrow [0, 1]$.
- $P(\mathbf{A}) = \prod_1^m P_i(A_i)$.

   where

- $\mathbf{A} = [A_1, ..., A_m]$,
- $A_i \in \chi_{Di}$,
- P_i is the probability measure defined on χ_{Di}.

Probability measures can be extended to *non-measurable events*, i.e., sets $A_j \in 2^D \setminus \chi_D$. Given D, χ_D and P, we define the *inner and outer extensions to P (P_* and P^*, respectively) as (Dudley 1989):

- $P_*, P^* : 2^D \rightarrow [0, 1]$
- $P_*(A) = sup\{P(X) : X \subseteq A, X \in \chi_D\} = P(\bigcup X : X \subseteq A, X \in \chi'_D)$
- $P^*(A) = inf\{P(X) : A \subseteq X, X \in \chi_D\} = P(\bigcup X : X \cap A \neq \{\}, X \in \chi'_D)$

Finally, inner and outer measures can be extended to cartesian products of a partition of D. Given a collection $D_1, ..., D_m$ of elements of a partition of D, and given also the algebras χ_{Di} and probability measures P_i of each $D_i, i = 1, ..., m$, we have:

- $P_{m*}, P_m^* : 2^{D_1 \times ... \times D_m} \rightarrow [0, 1]$
- $P_{m*}(A) = sup\{P_m(X) : X \subseteq A, X \in \chi\} = P_m(\bigcup X : X \subseteq A, X \in \chi')$.
- $P_m^*(A) = inf\{P_m(X) : A \subseteq X, X \in \chi\} = P_m(\bigcup X : X \cap A \neq \{\}, X \in \chi')$.

The measures P_{m*} and P_m^* can be regarded as approximations from below and from above to the probabilities of non-measurable events: if we could evaluate the probability $P_m(A)$, then we would have that $P_{m*}(A) \leq P_m(A) \leq P_m^*(A)$. Indeed, for measurable events we have that $P_{m*}(A) = P_m(A) = P_m^*(A)$.

As shown in (Fagin and Halpern 1989b), the best approximations we have for conditional probabilities of non-measurable events can be given by the following expressions:

$$\mathcal{P}_*(A|B) = \begin{cases} \frac{\mathcal{P}_*(A\cap B)}{\mathcal{P}_*(A\cap B)+\mathcal{P}^*(\neg A\cap B)}, & \mathcal{P}_*(A\cap B)+\mathcal{P}^*(\neg A\cap B)\neq 0 \\ 0, & \mathcal{P}_*(A\cap B)+\mathcal{P}^*(\neg A\cap B)=0 \end{cases}$$

$$\mathcal{P}^*(A|B) = \begin{cases} \frac{\mathcal{P}^*(A\cap B)}{\mathcal{P}^*(A\cap B)+\mathcal{P}_*(\neg A\cap B)}, & \mathcal{P}^*(A\cap B)+\mathcal{P}_*(\neg A\cap B)\neq 0 \\ 0, & \mathcal{P}^*(A\cap B)+\mathcal{P}_*(\neg A\cap B)=0 \end{cases}$$

For the case of measures on χ, these expressions can be stated as:

$$\mathcal{P}_{m*}(A|B) = \begin{cases} \frac{\mathcal{P}_{m*}(A\cap B)}{\mathcal{P}_{m*}(A\cap B)+\mathcal{P}^*_m(\neg A\cap B)}, & \mathcal{P}_{m*}(A\cap B)+\mathcal{P}^*_m(\neg A\cap B)\neq 0 \\ 0, & \mathcal{P}_{m*}(A\cap B)+\mathcal{P}^*_m(\neg A\cap B)=0 \end{cases}$$

$$\mathcal{P}^*_m(A|B) = \begin{cases} \frac{\mathcal{P}^*_m(A\cap B)}{\mathcal{P}^*_m(A\cap B)+\mathcal{P}_{m*}(\neg A\cap B)}, & \mathcal{P}^*_m(A\cap B)+\mathcal{P}_{m*}(\neg A\cap B)\neq 0 \\ 0, & \mathcal{P}^*_m(A\cap B)+\mathcal{P}_{m*}(\neg A\cap B)=0 \end{cases}$$

2.3 Probabilities of fuzzy events

A sentence containing vague predicates defines a fuzzy set of elements of the domain of discourse (or of elements of the cartesian product of members of one of the partitions of the domain of discourse). Hence, if we allow fuzzy predicates in our language, we must be prepared to specify the probability of fuzzy events.

In (Klement 1982) the concept of algebra is extended to fuzzy sets and in (Piasecki 1988, Smets 1982, Turksen 1988) the definition of the probability of a fuzzy event is presented, reputed as originally by L. Zadeh. A *fuzzy algebra* on D is defined by analogy with the concept of an algebra. It is a set χ_D^F of fuzzy subsets of D, such that (i) $\mu(A) = \text{constant} \Rightarrow A \in \chi_D^F$; (ii) $A \in \chi_D^F \Rightarrow \neg A \in \chi_D^F$; (iii) $A, B \in \chi_D^F \Rightarrow A \cup B \in \chi_D^F$.

Given an algebra χ_D and a probability measure $\mathcal{P}$ on χ_D, the probability of the fuzzy subset $A \in \chi_D^F$ is defined for every measurable A (i.e., for every $A \in \chi_D$), and is given by the Lebesgue-Stieltjes integral

$$\mathcal{P}^F(A) = \int_D \mu(A)d\mathcal{P}$$

From the computational point of view, we can access upper and lower bounds for this integral, related to the extreme values of the membership function in $A \cap A_i, A_i \in \chi'_D$:

$$\sum_{A_i \subseteq A} \mathcal{P}(A_i) \times min\{\mu(a_i): a_i \in A, \cap A\} \leq \mathcal{P}^F(A) \leq \sum_{A, \cap A \neq \{\}} \mathcal{P}(A_i) \times max\{\mu(a_i): a_i \in A, \cap A\}$$

These expressions can be extended to the non-measurable cases and to $2^{D_1 \times \ldots \times D_m}$, where $D_1, \ldots, D_m$ form a partition of D. Given a non-measurable fuzzy event A, we have:

$$\sum_{A_i \subseteq A} \mathcal{P}_m(A_i) \times min\{\mu(a_i): a_i \in A, \cap A\} \leq \mathcal{P}^F_m(A) \leq \sum_{A, \cap A \neq \{\}} \mathcal{P}_m(A_i) \times max\{\mu(a_i): a_i \in A, \cap A\}$$

And for the case of conditional probabilities, we have:

$$\mathcal{P}^F_{m*}(A|B) = \begin{cases} \frac{\mathcal{P}_{m*1}}{\mathcal{P}_{m*2}+\mathcal{P}^*_{m3}}, \mathcal{P}_{m*2}+\mathcal{P}^*_{m3}\neq 0 \\[2mm] 0, otherwise \end{cases}$$

$$\mathcal{P}^{F*}_m(A|B) = \begin{cases} min\{1,\frac{\mathcal{P}^*_{m1}}{\mathcal{P}^*_{m2}+\mathcal{P}_{m*3}}\}, \mathcal{P}^*_{m2}+\mathcal{P}_{m*3}\neq 0 \\[2mm] 0, otherwise \end{cases}$$

where

$$\mathcal{P}_{m*1} = \sum\nolimits_{A_i \subseteq [A\cap B]} \mathcal{P}_m(A_i) \times min\{\mu(a_i) : a_i \in A_i \cap [A \cap B]\}$$
$$\mathcal{P}_{m*2} = \sum\nolimits_{A_i \subseteq [A\cap B]} \mathcal{P}_m(A_i) \times max\{\mu(a_i) : a_i \in A_i \cap [A \cap B]\}$$
$$\mathcal{P}_{m*3} = \sum\nolimits_{A_i \subseteq [\neg A\cap B]} \mathcal{P}_m(A_i) \times min\{\mu(a_i) : a_i \in A_i \cap [\neg A \cap B]\}$$
$$\mathcal{P}^*_{m1} = \sum\nolimits_{A_i \cap [A\cap B]\neq \{\}} \mathcal{P}_m(A_i) \times max\{\mu(a_i) : a_i \in A_i \cap [A \cap B]\}$$
$$\mathcal{P}^*_{m2} = \sum\nolimits_{A_i \cap [A\cap B]\neq \{\}} \mathcal{P}_m(A_i) \times min\{\mu(a_i) : a_i \in A_i \cap [A \cap B]\}$$
$$\mathcal{P}^*_{m3} = \sum\nolimits_{A_i \cap [\neg A\cap B]\neq \{\}} \mathcal{P}_m(A_i) \times max\{\mu(a_i) : a_i \in A_i \cap [\neg A \cap B]\}$$

3 A Language Supporting Fuzzy Predicates

The relationship between fuzzy logics and the resolution principle is well established. Since (Lee 1972), one of the pioneering works in the area, several proposals have been made that aim at richer languages in respect of both the logical and the fuzzy relations supported.

In (Lee 1972) the language is limited to *definite clauses* (Apt 1987) *allowing fuzzy predicates with truth-values always greater than 0.5*. The semantics of the relevant connectives is defined according to Zadeh's triangular norms and conorms and resolution is extended to propagate truth-values in a way that is sound and complete with respect to the Herbrand interpretation of sets of clauses. Several implementations based on (Lee 1972) have been proposed, e.g., the ones described in (Hinde 1986, Ishizuka and Kanai 1985, Orci 1989).

More recent developments (Fitting 1988, Fitting 1990, Kifer and Subrahmanian 1991, Shapiro 1983, Van Emden 1986) have focused on fixpoint semantics, either working with definite programs or approaching the definition of negation by means other than finite failure. We adopt negation by finite failure here, in order to have the more conventional languages which are based on this principle (e.g., pure PROLOG) as proper subsets of our language. This choice is corroborated by the results found in (Turi 1989, Kunen 1989), which determine large classes of *normal programs* with a well-defined declarative semantics.

3.1 Reasoning with fuzzy predicates

The language presented here is defined after (Kunen 1989). The class of logic programs supported by this language is that of *function-free normal programs* under restrictions of *non-cyclicality, strictness with respect to queries*, and *allowedness* (see definitions of these terms in (Turi 1989, Kunen 1989, Corrêa da Silva 1992)), and its inference procedure - *SLDNF* - is known to be *sound* and *complete* with respect to the model of the *Clark's completion* of programs in the language (Kunen 1989):

Theorem 1. *Given a program P and a query ψ:*
  1. *$Comp(P) \models \psi$ iff ψ belongs to the success set of P ($\psi R\sigma$);*
  2. *$Comp(P) \models \neg\psi$ iff ψ belongs to the finite failure set of P (ψF).*

This defines a rich subset of first-order logic with a computationally efficient inference procedure and a formally specified declarative semantics.

A fuzzy predicate can be defined by analogy with the concept of fuzzy sets previously presented. The interpretation of predicates can be generalised to a function $I(p) : D^n \rightarrow [0, 1]$, with the extreme values corresponding to the previous values $\top$ and $\bot$ (namely, $\top \equiv 1$ and $\bot \equiv 0$). This function can be construed as a fuzzy membership function and the logical connectives can be interpreted as fuzzy set operators – '$\neg$' corresponding to complementation, '$\vee$' corresponding to union, '$\wedge$' corresponding to intersection, and '$\leftrightarrow$' corresponding to set-equivalence. Intuitively, the semantics of a closed formula becomes a "degree of truth", rather than simply one value out of $\{\top, \bot\}$. Let τ denote this value and $T(\psi, \tau)$ state that "the truth-degree of ψ is τ". This evaluation can be made operational using an *extended SLDNF (e-SLDNF)* procedure, to be related to the model of an *extended completion* of a program P ($e\text{-}Comp(P)$). We assume that the *unit clauses* (and only they) in the program express truth-degrees, that is, unit clauses are of the form $T(p, \tau)$, where $\tau > 0$.

The *extended completion* of a program P ($e\text{-}Comp\,(P)$) is defined as presented in Fig. 1.

A model for a program containing fuzzy predicates is any interpretation for which every expression φ occurring in $e\text{-}Comp(P)$ has a truth-value $\tau > 0$.

Two classes of formulae can be identified in $e\text{-}Comp(P)$:

- *unit formulae*, generated by rule 1 or from the unit clauses occurring in P; and
- *equivalence formulae*, i.e., the remaining ones, all of them containing the connective $\leftrightarrow$.

The connectives occurring in $e\text{-}Comp(P)$ are interpreted according to the truth-functions defined below:

  Assuming that:

- $T(\delta, \tau_\delta)$, and
- $T(\psi, \tau_\psi)$

  We have that:

- $T((\delta \wedge \psi), \tau) \Rightarrow \tau = min\{\tau_\delta, \tau_\psi\}$
- $T((\delta \vee \psi), \tau) \Rightarrow \tau = max\{\tau_\delta, \tau_\psi\}$
- $T((\neg\delta), \tau) \Rightarrow \tau = 1 - \tau_\delta$
- $\begin{cases} T((\delta \leftrightarrow \psi), 1) \Rightarrow \tau_\delta = \tau_\psi \\ T((\delta \leftrightarrow \psi), 0) \Rightarrow \tau_\delta \neq \tau_\psi \end{cases}$

Rules:

1. $$\frac{Def_p=\{\}}{\forall \mathbf{X}[T(\neg p(\mathbf{X}),1)]}$$

2. $$\frac{Def_p=\{p(\mathbf{t}_i)\leftarrow\psi_i::i=1,...,k\}\neq\{\}}{\forall \mathbf{X}[T(p(\mathbf{X}),\tau)\leftrightarrow max\{\tau_i::(\mathbf{X}=\mathbf{t}_i)}$$
$$\wedge[(\psi_i\neq\{\}\wedge T(\psi_i,\tau_i))$$
$$\vee(\psi_i=\{\}\wedge T(p(\mathbf{t}_i),\tau_i))]\}=\tau]$$

   where

   (a) Def_p is the set of clauses in P with p in the head;

   (b) $\mathbf{x}, \mathbf{t}_i$ are tuples of variables ($[x_1,...,x_m]$) and terms ($[t_{1i},...,t_{mi}]$), respectively;

   (c) $\mathbf{x}=\mathbf{t}_i$ stands for $x_1=t_{1i}\wedge...\wedge x_m=t_{mi}$;

   (d) ψ_i are (possibly empty) conjunctions of literals;

   (e) the connective $\leftrightarrow$ stands for equivalence.

Axioms:

1. equality axioms (Mendelson 1987);
2. $t(x)\neq x$ for each term in which x occurs.

Fig. 1. Extended completion of a program P

The completion of a conventional program defines a unique model for the program. For the extended completion to do the same, a necessary condition is to fix the truth-values for the unit clauses occurring in P as values greater than 0. This condition is also sufficient, as all the other formulae in $e\text{-}Comp(P)$ – i.e., the equivalence formulae and the unit formulae generated by rule 1 – must have truth-values equal to 1 in the model of the program.

Our notation for logic programs and the $e\text{-}SLDNF$ procedure is as follows:

- φ_i are literals;
- p_i, q_i are positive literals;
- g_i are positive ground literals;
- δ_i, ψ_i are (possibly empty) conjunctions of literals;
- σ, π are substitutions;
- R^e stands for "returns with a truth-value greater than 0": $\psi\, R^e\,(\sigma,\tau)$ holds iff $e\text{-}SLDNF$ succeeds, assigning a truth-value τ to ψ, with the substitution σ as an answer;
- F^e stands for "fails": ψF^e holds iff $e\text{-}SLDNF$ fails, implying on the assignment of a truth-value $\tau=0$ to ψ.
- $true$ stands for the empty query clause;
- yes stands for identity substitution.

$e\text{-}SLDNF$ is defined inductively as presented in Fig. 2.

1. $true \; R^e \; (yes, 1)$

2. $$\dfrac{(q,\delta),max\left\{\begin{array}{l}\tau_i:[p_i\leftarrow\psi_i],\sigma_i=mgu(q,p_i),(\psi_i,\delta)\sigma_i \; R^e \\ (\pi_i,\tau_i)\vee[p_i\leftarrow],\sigma_i=mgu(q,p_i),(\delta)\sigma_i \; R^e \\ (\pi_i,\tau_i'),T((p_i)\sigma_i,\tau_i''),min\{\tau_i',\tau_i''\}=\tau_i\end{array}\right\}=\tau}{(q,\delta) \; R^e \; (\sigma\pi,\tau)}$$

   where $\sigma\pi$ is the substitution that generates τ and the ψ_i are non-empty conjunctions.

3. $\dfrac{(\neg g,\delta),g \; R^e \; (yes,\tau'),\tau'\,\underset{\sim}{<}1,\delta \; R^e \; (\sigma,\tau''),min\{(1-\tau'),\tau''\}=\tau}{(\neg g,\delta) \; R^e \; (\sigma,\tau)}$

   $\dfrac{(\neg g,\delta),g F^e,\delta \; R^e \; (\sigma,\tau)}{(\neg g,\delta) \; R^e \; (\sigma,\tau)}$

4. $\dfrac{(q,\delta),\neg\exists[p\leftarrow\psi]:\exists mgu(q,p)}{(q,\delta) \; F^e}$

   $\dfrac{(q,\delta),\forall[p_i\leftarrow\psi_i]:\exists\sigma=mgu(q,p_i)\Rightarrow(\psi_i,\delta)\sigma \; F^e}{(q,\delta) \; F^e}$

5. $\dfrac{(\neg g,\delta),g \; R^e \; (yes,1)}{(\neg g,\delta) \; F^e}$

Fig. 2. *e-SLDNF*

4 A Language Supporting Probabilities on the Domain

The problem of representing and reasoning with statistical knowledge has received some attention recently (Bacchus 1990b, Halpern 1990). This problem can be roughly characterised as the problem of being able to represent in a first-order language terms of the form $\mathcal{P}_{\mathbf{x}}(\psi)$, to be read as "the probability of selecting a vector of instances for the variables in $\mathbf{x}$ that make ψ true".

In (Abadi and Halpern 1989) we have the result that the set of valid formulae for first-order logic containing statistical terms is not recursively enumerable, implying that a complete proof procedure for this logic does not exist. Two different ways of constraining the language to achieve proof-theoretic completeness have been proposed:

- in (Bacchus 1990b) the *probability measures* are relaxed to *non-σ-additive measures*, that is, the general probability axiom stating that "the probability of any *(infinitely) countable* set of pairwise disjoint events equals the sum of the probabilities of those events" is reduced to the case of *finite* sets of events. Moreover, the measures range on *real closed fields* (Shoenfield 1967) rather than on real numbers.
- in (Halpern 1990) the domain of discourse is *bounded in size*, i.e., it contains a number of elements not greater than a fixed N.

Our base language to be extended to contain statistical expressions obeys all these constraints: its domain is always finite and of fixed cardinality, so it is always bounded in size, σ-additivity coincides with finite additivity for finite domains and, as we intend to *compute* probabilities, field-valued measures are sufficient since, as pointed out in (Bacchus 1988), "computers are only capable of dealing with rational numbers (and only a finite set of them)".

On the other hand, the language introduced here extends the aforementioned results in two senses:

- we allow the occurrence of *fuzzy events*, i.e., statistical events that are characterised by fuzzy sets, and
- following a line suggested in (Halpern 1990), we admit the existence of *non-measurable events* and the consequent need for *inner and outer approximations* for statistical measures.

4.1 Reasoning with probabilities

Given a program P, the set of solutions with truth-values greater than 0 for a query ψ is always finite. This set also defines a fuzzy set of tuples of elements of D – the domain of P.

If our language is extended to accommodate the specification of probability measures of algebras of a partition of D through their bases, the set of solutions of ψ can be interpreted as a fuzzy event in the appropriate cartesian product of D, and upper and lower bounds can be evaluated for its probabilities using the measure of the corresponding cartesian product algebra.

The language is extended as follows:

- special unit formulae of the form $\mathcal{P}(S^c, \rho)$ are used to specify probability measures for D, i.e., a collection of expressions of the form $\mathcal{P}(S^c{}_{ij}, \rho_{ij})$ is attached to P, where the $S^c{}_{ij}$ form the bases of algebras χ_{Di} of a partition of D and ρ_{ij} is the probability of $S^c{}_{ij}$;
- some definitions are implicitly assumed as part of our inference procedure: the definitions of the operations of *addition* $(+)$ and *multiplication* $(\times)$, of the *relations* $>$ and $=$, and of the properties of *non-negativity* $(\rho \geq 0 \leftarrow \mathcal{P}(S^c, \rho))$, finite additivity $(\mathcal{P}(\cup_1^n S^c{}_i, \rho) \leftarrow \mathcal{P}(S^c{}_1, \rho_1), ..., \mathcal{P}(S^c{}_n, \rho_n), \rho = \rho_1 + ... + \rho_n)$ and total probability $(\mathcal{P}(D, 1))$.
- special second-order expressions of the forms $\mathcal{P}_*(S, \psi, \rho_*)$ and $\mathcal{P}^*(S, \psi, \rho^*)$ are introduced, to be read as "the lower and the upper bounds for the probability of having a tuple of instances for the variables in S which satisfies ψ are ρ_* and ρ^*".
- special second-order expressions of the forms $\mathcal{P}_*(S, \psi_1 | \psi_2, \rho_*)$ and $\mathcal{P}^*(S, \psi_1 | \psi_2, \rho^*)$ are introduced, to be read as "the lower and the upper bounds for the probability of having a tuple of instances for the variables in S which satisfies ψ_1 given ψ_2 are ρ_* and ρ^*".

$\mathcal{P}_*(S, \psi, \rho_*)$ and $\mathcal{P}^*(S, \psi, \rho^*)$ are evaluated as follows:

1. generate K_ψ, the finite fuzzy set of tuples of instances of the free variables in ψ which associate a non-zero truth-degree τ to ψ.

If ψ does not have free variables, K_ψ is the singleton set containing the tuple of terms occurring in ψ with its respective truth-degree.

If ψ contains free variables, then K_ψ is generated by substituting exhaustively each free variable in ψ by elements of D and then selecting the substitutions which generate the desired truth-degrees.

2. generate K_ψ^S - the projection of K_ψ over S: select from the tuple of free variables in ψ those which are also in S, and extract the corresponding tuples of instances from K_ψ.

3. If $K_\psi^S \neq \{\}$ then:

(a) generate the cartesian product algebra and measure of the same arity as the tuples in K_ψ^S assuming the elements of D to be statistically independent.

(b) generate ρ_* and ρ^*:

$$\rho_* = \sum_{S_i \subseteq K_\psi^S} \mathcal{P}_m(S_i) \times min\{\tau : \mu(\mathbf{K}) = \tau, \mathbf{K} \in S_i \cap K_\psi^S\}, S_i \in \chi_D'$$

$$\rho^* = \sum_{S_i \cap K_\psi^S \neq \{\}} \mathcal{P}_m(S_i) \times max\{\tau : \mu(\mathbf{K}) = \tau, \mathbf{K} \in S_i \cap K_\psi^S\}, S_i \in \chi_D'$$

where m is the arity of the tuples in K_ψ^S.

If $K_\psi^S = \{\}$ then make $\rho_* = \rho^* = \tau$, where $\mathcal{T}(\psi, \tau)$.

$\mathcal{P}_*(S, \psi_1|\psi_2, \rho_*)$ and $\mathcal{P}^*(S, \psi_1|\psi_2, \rho^*)$ are evaluated as follows:

1. generate $K_{(\psi_1,\psi_2)}^S$ and $K_{(\neg\psi_1,\psi_2)}^S$.

2. generate $min_\wedge, min_\wedge', min_\wedge^-, max_\wedge, max_\wedge'$ and $max_\wedge^-$:

$$min_\wedge = \sum_{S_i \subseteq K_{(\psi_1,\psi_2)}^S} \mathcal{P}_m(S_i) \times min\{\tau : \mu(\mathbf{K}) = \tau, \mathbf{K} \in S_i \cap K_{(\psi_1,\psi_2)}^S\}, S_i \in \chi_D'$$

$$min_\wedge' = \sum_{S_i \cap K_{(\psi_1,\psi_2)}^S \neq \{\}} \mathcal{P}_m(S_i) \times min\{\tau : \mu(\mathbf{K}) = \tau, \mathbf{K} \in S_i \cap K_{(\psi_1,\psi_2)}^S\}, S_i \in \chi_D'$$

$$min_\wedge^- = \sum_{S_i \subseteq K_{(\neg\psi_1,\psi_2)}^S} \mathcal{P}_m(S_i) \times min\{\tau : \mu(\mathbf{K}) = \tau, \mathbf{K} \in S_i \cap K_{(\neg\psi_1,\psi_2)}^S\}, S_i \in \chi_D'$$

$$max_\wedge = \sum_{S_i \cap K_{(\psi_1,\psi_2)}^S \neq \{\}} \mathcal{P}_m(S_i) \times max\{\tau : \mu(\mathbf{K}) = \tau, \mathbf{K} \in S_i \cap K_{(\psi_1,\psi_2)}^S\}, S_i \in \chi_D'$$

$$max_\wedge' = \sum_{S_i \subseteq K_{(\psi_1,\psi_2)}^S} \mathcal{P}_m(S_i) \times max\{\tau : \mu(\mathbf{K}) = \tau, \mathbf{K} \in S_i \cap K_{(\psi_1,\psi_2)}^S\}, S_i \in \chi_D'$$

$$max_\wedge^- = \sum_{S_i \cap K_{(\neg\psi_1,\psi_2)}^S \neq \{\}} \mathcal{P}_m(S_i) \times max\{\tau : \mu(\mathbf{K}) = \tau, \mathbf{K} \in S_i \cap K_{(\neg\psi_1,\psi_2)}^S\}, S_i \in \chi_D'$$

3. generate ρ_* and ρ^*:

$$\rho_* = \begin{cases} \frac{min_\wedge}{max_\wedge' + max_\wedge^-}, & max_\wedge' + max_\wedge^- \neq 0 \\ 0, & max_\wedge' + max_\wedge^- = 0 \end{cases}$$

$$\rho^* = \begin{cases} min\{1, \frac{max_\wedge}{min_\wedge' + min_\wedge^-}\}, & min_\wedge' + min_\wedge^- \neq 0 \\ 0, & min_\wedge' + min_\wedge^- = 0 \end{cases}$$

Since probabilities are completely defined by measures on the constants of the language, terms of the forms $\mathcal{P}_*(S, \psi, \rho_*)$, $\mathcal{P}^*(S, \psi, \rho^*)$, $\mathcal{P}_*(S, \psi_1|\psi_2, \rho_*)$ and $\mathcal{P}^*(S, \psi_1|\psi_2, \rho^*)$ never occur as heads of program clauses. Moreover, these terms only admit truth-degrees in $\{0, 1\}$.

5 A Language Supporting Degrees of Belief

5.1 Adding possible worlds

The concept of *possible worlds* has been evoked frequently as a useful device to aid modelling uncertainty (see for example (Bacchus 1990b, Bundy 1985, Fagin and Halpern 1989a, Nilsson 1986)). The general idea is the assumption that there is a collection of *worlds* (or *states*, or *interpretations*), each of them assigning different truth-values to the formulae in our language. Intuitively, a possible world should be viewed as a conceivable hypothetical scenario upon which we can construct our reasoning.

Given a program P and a set of possible worlds $\Omega = \{\omega_1, ...\}$, a *rigid formula* is a formula which is always assigned the same truth-value in all possible worlds.

We assume in our language that, given a program P, each possible world ω_i assigns a different fuzzy truth-value to the set of unit clauses in P. We assume that the other clauses occurring in P, i.e., that the logical dependency and statistical relations expressed in P, are rigid.

Ideally, we should keep track of every possible world independently, and repeatedly apply the machinery presented in the previous sections for each of them each time we activated P with a query ψ. This procedure becomes computationally intractable as the size of Ω gets bigger (notice that Ω is not even required to be finite. Obviously it would not be possible to keep track computationally of an infinite set of possible worlds). Alternatively, we should be able to calculate singular truth-values like the minimum and the maximum values occurring in Ω for each clause: given a program P with unit clauses of the form $T_*(C_i, \tau_{*i})$ and $T^*(C_i, \tau_i^*)$ (representing minimum and maximum truth-degrees, respectively), we should be able to derive the values $T_*(\psi, \tau_*)$ and $T^*(\psi, \tau^*)$ for a query ψ.

It is not possible, however, to obtain these values for any query given only the singular values for the unit clauses, as the example below shows:

Example 1. Consider the following program:

$r(a) \leftarrow p(a), q(a).$
$s(a) \leftarrow p(a).$
$s(a) \leftarrow q(a).$
$T_*(p(a), 0.2).\ T^*(p(a), 0.8).$
$T_*(q(a), 0.3).\ T^*(q(a), 0.6).$

Assume further that the truth-degrees have come from the possible worlds in Ω, according to Table 1 (values underlined):

$\Omega = \{\omega_1, \omega_2, \omega_3\}$

	ω_1	ω_2	ω_3
$T(p(a))$	0.2	0.5	0.8
$T(q(a))$	0.4	0.6	0.3

Table 1. Truth-degrees in Ω

Using the procedure introduced in Fig. 2 for each possible world separately, the results in Table 2 follow:

	ω_1	ω_2	ω_3
$\mathcal{T}(r(a))$	0.2	0.5	0.3
$\mathcal{T}(s(a))$	0.4	0.6	0.8

Table 2. Derived Truth-degrees in Ω

which indicate that (values underlined, Table 2):

$\mathcal{T}_*(r(a), 0.2)$. $\mathcal{T}^*(r(a), 0.5)$.
$\mathcal{T}_*(s(a), 0.4)$. $\mathcal{T}^*(s(a), 0.8)$.

Both $\mathcal{T}^*(r(a), 0.5)$ and $\mathcal{T}_*(s(a), 0.4)$ depend on information which is not given in the initial program, thus cannot be derived from it unless we use additional knowledge.

Approximate solutions can be obtained for $\mathcal{T}_*$ and $\mathcal{T}^*$, i.e., we can obtain the values $\hat{\mathcal{T}}_*$ and $\hat{\mathcal{T}}^*$, such that $\hat{\mathcal{T}}_*(\psi) \leq \mathcal{T}_*(\psi)$ and $\hat{\mathcal{T}}^*(\psi) \geq \mathcal{T}^*(\psi)$ for any query ψ.

It is not difficult to verify that the following recursive rules satisfy these conditions:

$$r \leftarrow p \Rightarrow \hat{\mathcal{T}}_*(r) = \hat{\mathcal{T}}_*(p)$$
$$\hat{\mathcal{T}}^*(r) = \hat{\mathcal{T}}^*(p)$$

$$r \leftarrow P, Q \Rightarrow \hat{\mathcal{T}}_*(r) = min\{\hat{\mathcal{T}}_*(P), \hat{\mathcal{T}}_*(Q)\}$$
$$\hat{\mathcal{T}}^*(r) = min\{\hat{\mathcal{T}}^*(P), \hat{\mathcal{T}}^*(Q)\}$$

$$r \leftarrow P \Rightarrow \hat{\mathcal{T}}_*(r) = max\{\hat{\mathcal{T}}_*(P), \hat{\mathcal{T}}_*(Q)\}$$
$$r \leftarrow Q \quad \hat{\mathcal{T}}^*(r) = max\{\hat{\mathcal{T}}^*(P), \hat{\mathcal{T}}^*(Q)\}$$

$$r \leftarrow \neg p \Rightarrow \hat{\mathcal{T}}_*(r) = 1 - \hat{\mathcal{T}}^*(p)$$
$$\hat{\mathcal{T}}^*(r) = 1 - \hat{\mathcal{T}}_*(p)$$

where $p, q, r, \ldots$ denote atoms and $P, Q, R, \ldots$ denote conjunctions of literals. When applied to the example above, these rules give:

$\hat{\mathcal{T}}_*(r(a), 0.2)$. $\hat{\mathcal{T}}^*(r(a), 0.6)$.
$\hat{\mathcal{T}}_*(s(a), 0.3)$. $\hat{\mathcal{T}}^*(s(a), 0.8)$.

which obey the desired inequality conditions.

5.2 Reasoning with possible worlds

The *e-SLDNF* procedure and the completion *e-Comp*(P) presented in Section 3 must be changed to accommodate the bounds for the truth-degrees across possible worlds. The *completion* of P is redefined as $*$-$Comp(P)$ as presented in Fig. 3.

Rules:

1. $$\frac{Def_p=\{\}}{\begin{array}{l}\forall X[\mathcal{T}_*(\neg p(X),1)]\\\forall X[\mathcal{T}^*(\neg p(X),1)]\end{array}}$$

2. $$\frac{Def_p=\{p(t_i)\leftarrow\psi_i:i=1,\dots,k\}\neq\{\}}{\begin{array}{l}\forall X[\mathcal{T}_*(p(X),\tau_*)\leftarrow max\{\tau_{*i}:(X=t_i)\wedge[(\psi_i\neq\{\},\mathcal{T}_*(\psi_i,\tau_{*i}))\vee\\(\psi_i=\{\},\mathcal{T}_*(p(t_i),\tau_{*i}))]\}=\tau_*]\end{array}}$$

$$\forall X[\mathcal{T}^*(p(X),\tau^*)\leftarrow max\{\tau_i^*:(X=t_i)\wedge[(\psi_i\neq\{\},\mathcal{T}^*(\psi_i,\tau_i^*))$$
$$\vee(\psi_i=\{\},\mathcal{T}^*(p(t_i),\tau_i^*))]\}=\tau^*]$$

where
   (a) Def_p is the set of clauses in P with p in the head;
   (b) x, t_i are tuples of variables $([x_1, ..., x_m])$ and terms $([t_{1i}, ..., t_{mi}])$, respectively;
   (c) $x = t_i$ stands for $x_1 = t_{1i} \wedge ... \wedge x_m = t_{mi}$;
   (d) the scope of the existential quantifier are the variables occurring in the bodies of the clauses in Def_p;
   (e) ψ_i are (possibly empty) conjunctions of literals;
   (f) the connective $\leftrightarrow$ stands for equivalence.

Axioms: the same as in Fig. 1.

Fig. 3. $*$-$Comp(P)$

A model for a program P is any interpretation for which every expression occurring in $*$-$Comp(P)$ has a value $\tau^* > 0$.

In order to redefine the inference procedure as $*$-$SLDNF$, we need the following in our notation for the success and finite failure set:

- R_*, R^* :
  $\psi R_*(\psi, \tau_*)$ holds iff $*$-$SLDNF$ succeeds, assigning τ_* to ψ as a lower bound for its truth-degree;
  $\psi R^*(\psi, \tau^*)$ holds iff $*$-$SLDNF$ succeeds, assigning τ^* to ψ as an upper bound for its truth-degree.
- F_*, F^* :
  ψF_* holds iff $*$-$SLDNF$ fails, assigning $\tau_* = 0$ to ψ;
  ψF^* holds iff $*$-$SLDNF$ fails, assigning $\tau^* = 0$ to ψ.

$*$-$SLDNF$ is defined inductively as presented in Fig. 4.

This language subsumes the one presented in the previous section, having that language as the particular class of programs in which all truth-degrees are rigid.

1. $true\ R_*\ (yes, 1)$.
   $true\ R^*\ (yes, 1)$.

2.
$$(q,\delta), max \begin{cases} \tau_{*i} : [p_i \leftarrow \psi_i], \sigma_i = mgu(q,p_i), \\ (\psi_i, \delta)\sigma_i\ R_*\ (\pi_i, \tau_{*i}) \vee \\ [p_i \leftarrow], \sigma_i = mgu(q,p_i), (\delta)\sigma_i\ R_*\ (\pi_i, \tau_i'), \\ \mathcal{T}_*((p_i)\sigma_i, \tau_i''), min\{\tau_i', \tau_i''\} = \tau_{*i} \end{cases} = \tau_* $$
$$\overline{\qquad\qquad\qquad (q,\delta)\ R_*\ (\sigma\pi, \tau_*) \qquad\qquad\qquad}$$

$$(q,\delta), max \begin{cases} \tau_i^* : [p_i \leftarrow \psi_i], \sigma_i = mgu(q,p_i), \\ (\psi_i, \delta)\sigma_i\ R^*\ (\pi_i, \tau_i^*) \vee \\ [p_i \leftarrow], \sigma_i = mgu(q,p_i), (\delta)\sigma_i\ R^*\ (\pi_i, \tau_i'), \\ \mathcal{T}^*((p_i)\sigma_i, \tau_i''), min\{\tau_i', \tau_i''\} = \tau_i^* \end{cases} = \tau^* $$
$$\overline{\qquad\qquad\qquad (q,\delta)\ R^*\ (\sigma\pi, \tau^*) \qquad\qquad\qquad}$$

where $\sigma\pi$ are the substitutions that generate τ_* and τ^* and the ψ_i are non-empty conjunctions.

$$\frac{(\neg g, \delta), g\ R^*\ (yes, \tau'), \tau' < 1, \delta\ R_*\ (\sigma, \tau''), min\{(1-\tau'), \tau''\} = \tau_*}{(\neg g, \delta)\ R_*\ (\sigma, \tau_*)}$$

$$\frac{(\neg g, \delta), g\ R_*\ (yes, \tau'), \tau' < 1, \delta\ R^*\ (\sigma, \tau''), min\{(1-\tau'), \tau''\} = \tau^*}{(\neg g, \delta)\ R^*\ (\sigma, \tau^*)}$$

3.
$$\frac{(\neg g, \delta), g F^*, \delta\ R_*\ (\sigma, \tau_*)}{(\neg g, \delta)\ R_*\ (\sigma, \tau_*)}$$

$$\frac{(\neg g, \delta), g F_*, \delta\ R^*\ (\sigma, \tau^*)}{(\neg g, \delta)\ R^*\ (\sigma, \tau^*)}$$

4.
$$\frac{(q, \delta), \neg \exists [p \rightarrow \psi] : \exists mgu(q,p)}{(q, \delta)\ F_*}$$

$$\frac{(q, \delta), \neg \exists [p \rightarrow \psi] : \exists mgu(q,p)}{(q, \delta)\ F^*}$$

$$\frac{(q, \delta), \forall [p_i \leftarrow \psi_i] : \exists \sigma = mgu(q,p_i) \Rightarrow (\psi_i, \delta)\sigma\ F_*}{(q, \delta)\ F_*}$$

$$\frac{(q, \delta), \forall [p_i \leftarrow \psi_i] : \exists \sigma = mgu(q,p_i) \Rightarrow (\psi_i, \delta)\sigma\ F^*}{(q, \delta)\ F^*}$$

5.
$$\frac{(\neg g, \delta), g\ R^*\ (yes, 1)}{(\neg g, \delta)\ F_*}$$

$$\frac{(\neg g, \delta), g\ R_*\ (yes, 1)}{(\neg g, \delta)\ F^*}$$

Fig. 4. **-SLDNF*

5.3 Probabilities on the domain with possible worlds

Probability evaluations take into account the bounds for truth-degrees across possible worlds. The syntax of the language can be as before for declaring probabilities, but the evaluation procedure must be changed. In the case of $\mathcal{P}_*(S, \psi, \rho_*)$ we must:

1. generate $K_{\psi*}$, the finite fuzzy set of tuples of instances of the free variables in ψ which associate a non-zero lower bound for the truth-degree τ_* to ψ.

   If ψ does not have free variables, $K_{\psi*}$ is the singleton set containing the tuple of terms occurring in ψ with its respective bound for the truth-degree.

   If ψ contains free variables, then $K_{\psi*}$ is generated by substituting exhaustively each free variable in ψ by elements of D and then selecting the substitutions which generate the desired bounds for truth-degrees.

2. generate $K_{\psi*}^S$ – the projection of $K_{\psi*}$ over S: select from the tuple of free variables in ψ those which are also in S, and extract the corresponding tuples of instances from $K_{\psi*}$.

3. case 1: $K^S_{\psi *} \neq \{\}$

(a) generate the cartesian product algebra and measure of the same arity as the tuples in $K^S_{\psi *}$ assuming the elements of D to be statistically independent.

(b) generate ρ_*:

$$\rho_* = \sum\nolimits_{\mathbf{S}_i \subseteq K^S_{\psi *}} \mathcal{P}_m(\mathbf{S}_i) \times min\{\tau_* : \mu_*(\mathbf{K}) = \tau_*, \mathbf{K} \in \mathbf{S}_i \cap K^S_{\psi *}\}, \mathbf{S}_i \in \chi'_D$$

case 2: $K^S_{\psi *} = \{\}$:

make $\rho_* = \tau_*$, where $\hat{T}_*(\psi, \tau_*)$.

In the case of $\mathcal{P}^*(S, \psi, \rho^*)$, we must:

1. generate K^*_ψ, the finite fuzzy set of tuples of instances of the free variables in ψ which associate a non-zero upper bound for the truth-degree τ^* to ψ.

If ψ does not have free variables, K^*_ψ is the singleton set containing the tuple of terms occurring in ψ with its respective bound for the truth-degree.

If ψ contains free variables, then K^*_ψ is generated by substituting exhaustively each free variable in ψ by elements of D and then selecting the substitutions which generate the desired bounds for truth-degrees.

2. generate K^{*S}_ψ – the projection of K^*_ψ over S: select from the tuple of free variables in ψ those which are also in S, and extract the corresponding tuples of instances from K^*_ψ.

3. case 1: $K^{*S}_\psi \neq \{\}$

(a) generate the cartesian product algebra and measure of the same arity as the tuples in K^{*S}_ψ assuming the elements of D to be statistically independent.

(b) generate ρ^*:

$$\rho^* = \sum\nolimits_{\mathbf{S}_i \cap K^{*S}_\psi \neq \{\}} \mathcal{P}_m(\mathbf{S}_i) \times max\{\tau^* : \mu^*(\mathbf{K}) = \tau^*, \mathbf{K} \in \mathbf{S}_i \cap K^{*S}_\psi\}, \mathbf{S}_i \in \chi'_D$$

case 2: $K^{*S}_\psi = \{\}$:

make $\rho^* = \tau^*$, where $\hat{T}^*(\psi, \tau^*)$.

In the cases of $\mathcal{P}_*(S, \psi_1|\psi_2, \rho_*)$ and $\mathcal{P}^*(S, \psi_1|\psi_2, \rho^*)$, we must:

1. generate $K^{*S}_{(\psi_1, \psi_2)}$, $K^{*S}_{(\neg\psi_1, \psi_2)}$, $K^S_{(\psi_1, \psi_2)*}$ and $K^S_{(\neg\psi_1, \psi_2)*}$.

2. generate $min_\wedge, min'_\wedge, min^-_\wedge, max_\wedge, max'_\wedge$ and $max^-_\wedge$:

$$min_\wedge = \sum\nolimits_{\mathbf{S}_i \subseteq K^S_{(\psi_1, \psi_2)*}} \mathcal{P}_m(\mathbf{S}_i) \times min\{\tau_* : \mu_*(\mathbf{K}) = \tau_*, \mathbf{K} \in \mathbf{S}_i \cap K^S_{(\psi_1, \psi_2)*}\}, \mathbf{S}_i \in \chi'_D$$

$$min'_\wedge = \sum\nolimits_{\mathbf{S}_i \cap K^{*S}_{(\psi_1, \psi_2)} \neq \{\}} \mathcal{P}_m(\mathbf{S}_i) \times min\{\tau_* : \mu_*(\mathbf{K}) = \tau_*, \mathbf{K} \in \mathbf{S}_i \cap K^{*S}_{(\psi_1, \psi_2)}\}, \mathbf{S}_i \in \chi'_D$$

$$min^-_\wedge = \sum\nolimits_{\mathbf{S}_i \subseteq K^S_{(\neg\psi_1, \psi_2)*}} \mathcal{P}_m(\mathbf{S}_i) \times min\{\tau_* : \mu_*(\mathbf{K}) = \tau_*, \mathbf{K} \in \mathbf{S}_i \cap K^S_{(\neg\psi_1, \psi_2)*}\}, \mathbf{S}_i \in \chi'_D$$

$$max_\wedge = \sum\nolimits_{\mathbf{S}_i \cap K^{*S}_{(\psi_1, \psi_2)} \neq \{\}} \mathcal{P}_m(\mathbf{S}_i) \times max\{\tau^* : \mu^*(\mathbf{K}) = \tau^*, \mathbf{K} \in \mathbf{S}_i \cap K^{*S}_{(\psi_1, \psi_2)}\}, \mathbf{S}_i \in \chi'_D$$

$$max'_\wedge = \sum\nolimits_{\mathbf{S}_i \subseteq K^S_{(\psi_1, \psi_2)*}} \mathcal{P}_m(\mathbf{S}_i) \times max\{\tau^* : \mu^*(\mathbf{K}) = \tau^*, \mathbf{K} \in \mathbf{S}_i \cap K^S_{(\psi_1, \psi_2)*}\}, \mathbf{S}_i \in \chi'_D$$

$$max^-_\wedge = \sum\nolimits_{\mathbf{S}_i \cap K^{*S}_{(\neg\psi_1, \psi_2)} \neq \{\}} \mathcal{P}_m(\mathbf{S}_i) \times max\{\tau^* : \mu^*(\mathbf{K}) = \tau^*, \mathbf{K} \in \mathbf{S}_i \cap K^{*S}_{(\neg\psi_1, \psi_2)}\}, \mathbf{S}_i \in \chi'_D$$

3. generate ρ_* and ρ^* :

$$\rho_* = \begin{cases} \frac{min_\wedge}{max'_\wedge + max_\wedge^\neg}, & max'_\wedge + max_\wedge^\neg \neq 0 \\ 0, & max'_\wedge + max_\wedge^\neg = 0 \end{cases}$$

$$\rho^* = \begin{cases} min\{1, \frac{max_\wedge}{min'_\wedge + min_\wedge^\neg}\}, & min'_\wedge + min_\wedge^\neg \neq 0 \\ 0, & min'_\wedge + min_\wedge^\neg = 0 \end{cases}$$

6 Adding Probabilities on Possible Worlds

Different worlds can have different likelihoods. Given a set of possible worlds Ω, we can define a *probability measure* $\mathcal{B}$ to represent these likelihoods. The *expected value* for the truth-degree of a clause ψ can be defined as $\mathcal{B}(\psi) = \int_\Omega \tau d\beta$, where $\mathcal{T}(\psi, \tau)$.

When the sets ω of possible worlds in which sentences have "non-zero" truth-degrees are measurable, this defines a straightforward extension of Nilsson's probabilistic logic (Nilsson 1986) to deal with fuzzy predicates. If we consider the non-measurable cases, then the language extends the so-called *Dempster-Shafer structures* (Fagin and Halpern 1989a, Corrêa da Silva and Bundy 1990), which are expressive enough to represent what other important mechanisms to represent degrees of belief can represent, such as Dempster-Shafer Belief and Plausibility Measures, Possibilistic Logic (Dubois and Prade 1988, Ruspini 1989) and Incidence Calculus (Bundy 1985, Corrêa da Silva and Bundy 1991).

Given expressions of the form $\mathcal{B}_*(\psi, \beta_*)$ and $\mathcal{B}^*(\psi, \beta^*)$ for the unit clauses occurring in a program P – where β_* and β^* represent the *inner* and the *outer extensions* for the measure $\mathcal{B}$ – it is not possible to derive the degrees of belief for all queries on P, unless the statistical dependency among clauses is known (Bundy 1985). Nonetheless, *bounds* can be derived for these degrees of belief. For programs without fuzzy predicates, these bounds can be defined by the following rules (from (Ng and Subrahmanian 1992)):

$$r \leftarrow p \Rightarrow \mathcal{B}_*(r) = \mathcal{B}_*(p)$$
$$\mathcal{B}^*(r) = \mathcal{B}^*(p)$$

$$r \leftarrow P, Q \Rightarrow \mathcal{B}_*(r) \geq max\{0, [\mathcal{B}_*(P) + \mathcal{B}_*(Q) - 1]\}$$
$$\mathcal{B}^*(r) \leq min\{\mathcal{B}^*(P), \mathcal{B}^*(Q)\}$$

$$r \leftarrow P \Rightarrow \mathcal{B}_*(r) \geq max\{\mathcal{B}_*(P), \mathcal{B}_*(Q)\}$$
$$r \leftarrow Q \quad \mathcal{B}^*(r) \leq min\{1, [\mathcal{B}^*(P) + \mathcal{B}^*(Q)]\}$$

$$r \leftarrow \neg p \Rightarrow \mathcal{B}_*(r) = 1 - \mathcal{B}^*(p)$$
$$\mathcal{B}^*(r) = 1 - \mathcal{B}_*(p)$$

where $p, q, r, \ldots$ denote atoms and $P, Q, R, \ldots$ denote conjunctions of literals.

When a program contains fuzzy information, these rules can be further refined, since we have that $\hat{\mathcal{T}}_*(\psi) \leq \mathcal{B}(\psi) \leq \hat{\mathcal{T}}^*(\psi)$ for any clause ψ:

$$r \leftarrow p \Rightarrow \mathcal{B}_{*}(r) = \mathcal{B}_{*}(p)$$
$$\mathcal{B}^{*}(r) = \mathcal{B}^{*}(p)$$

$$r \leftarrow P, Q \Rightarrow \mathcal{B}_{*}(r) \geq \mathcal{B}_{*}(r) = max\{\mathcal{T}_{*}(r), [\mathcal{B}_{*}(P) + \mathcal{B}_{*}(Q) - 1]\}$$
$$\mathcal{B}^{*}(r) \leq \mathcal{B}^{*}(r) = min\{\mathcal{T}^{*}(r), \mathcal{B}^{*}(P), \mathcal{B}^{*}(Q)\}$$

$$r \leftarrow P \Rightarrow \mathcal{B}_{*}(r) \geq \mathcal{B}_{*}(r) = max\{\mathcal{T}_{*}(r), \mathcal{B}_{*}(P), \mathcal{B}_{*}(Q)\}$$
$$r \leftarrow Q \quad \mathcal{B}^{*}(r) \leq \mathcal{B}^{*}(r) = min\{\mathcal{T}^{*}(r), [\mathcal{B}^{*}(P) + \mathcal{B}^{*}(Q)]\}$$

$$r \leftarrow \neg p \Rightarrow \mathcal{B}_{*}(r) = 1 - \mathcal{B}^{*}(p)$$
$$\mathcal{B}^{*}(r) = 1 - \mathcal{B}_{*}(p)$$

In what follows we introduce these concepts into our language.

6.1 Reasoning with probabilities on possible worlds

The inference procedure defined in the previous sections can be extended to deal with degrees of belief. We define, in addition to $*$-$SLDNF$ and $*$-$Comp(P)$ previously presented, the procedure and completion rules for evaluating degrees of belief in a program P presented in Fig. 5 and Fig. 6, in which the following notational conventions are adopted:

- $\psi R_{\beta_{*}}(\psi, \beta_{*})$ holds iff $*$-$SLDNF$ succeeds, assigning β_{*} to ψ as a lower bound for its truth-degree;
- $\psi R_{\beta}^{*}(\psi, \beta^{*})$ holds iff $*$-$SLDNF$ succeeds, assigning β^{*} to ψ as an upper bound for its truth-degree;
- $\psi F_{\beta_{*}}$ holds iff $*$-$SLDNF$ fails, assigning $\beta_{*} = 0$ to ψ;
- ψF_{β}^{*} holds iff $*$-$SLDNF$ fails, assigning $\beta^{*} = 0$ to ψ.

Rules:

1. $$\frac{Def_{p} = \{\}}{\forall X[\mathcal{B}_{*}(\neg p(X), 1)]}$$
$$\forall X[\mathcal{B}^{*}(\neg p(X), 1)]$$

2. $$\frac{Def_{p} = \{p(t_{i}) \leftarrow \psi_{i} :: i = 1, ..., k\} \neq \{\}}{\forall X[\mathcal{B}_{*}(p(X), \beta_{*}) \leftarrow max\{\beta_{*_{i}} :: (X = t_{i}) \wedge [(\psi_{i} \neq \{\}, \mathcal{B}_{*}(\psi_{i}, \beta_{*_{i}}))}$$
$$\vee (\psi_{i} = \{\}, \mathcal{B}_{*}(p(t_{i}), \beta_{*_{i}}))]\} = \mathcal{B}_{*}, \mathcal{T}_{*}(p(X, r_{*}), max\{\mathcal{B}_{*}, r_{*}\} = \beta_{*}]$$

$$\forall X[\mathcal{B}^{*}(p(X), \beta^{*}) \leftarrow \sum \{\beta_{i}^{*} :: (X = t_{i}) \wedge [(\psi_{i} \neq \{\}, \mathcal{B}^{*}(\psi_{i}, \beta_{i}^{*}))$$
$$\vee (\psi_{i} = \{\}, \mathcal{B}^{*}(p(t_{i}), \beta_{i}^{*}))]\} = \mathcal{B}^{*}, \mathcal{T}^{*}(p(X, \tau^{*}), min\{\mathcal{B}^{*}, \tau^{*}\} = \beta^{*}]$$

Axioms: the same as in Fig. 1.

Fig. 5. β-$Comp(P)$

1. $true\ R_{\beta_\bullet}\ (yes, 1)$.
   $true\ R_{\beta}^*\ (yes, 1)$.

$$max_{(q,\delta),} \left\{ \begin{array}{l} \beta_{\bullet i}:[p_i \leftarrow \psi_i], \sigma_i = mgu(q, p_i), \\ (\psi_i, \delta)1z\sigma_i\ R_{\beta_\bullet}\ (\pi_i, \beta_{\bullet i}) \vee [p_i \leftarrow], \\ \sigma_i = mgu(q, p_i), (\delta)\sigma_i\ R_{\beta_\bullet}\ (\pi_i, \beta_i'), \\ \beta_\bullet(p_i\sigma_i, \beta_i''), \beta_{\bullet i} = \beta_i' + \beta_i'' - 1 \end{array} \right\} =$$

2. $$\dfrac{\begin{array}{c} \beta_\bullet', \mathcal{T}_\bullet((q,\delta), \tau_\bullet), \\ max\{\tau_\bullet, \beta_\bullet'\} = \beta_\bullet \end{array}}{(q,\delta)\ R_{\beta_\bullet}\ (\sigma\pi, \beta_\bullet)}.$$

$$\sum_{(q,\delta),} \left\{ \begin{array}{l} \beta_i^*:[p_i \leftarrow \psi_i], \sigma_i = mgu(q, p_i), \\ (\psi_i, \delta)\sigma_i\ R_\beta^*\ (\pi_i, \beta_i^*) \vee [p_i \leftarrow], \\ \sigma_i = mgu(q, p_i), (\delta)\sigma_i\ R_\beta^*\ (\pi_i, \beta_i'), \\ \beta^*(p_i\sigma_i, \beta_i''), min\{\beta_i', \beta_i''\} = \beta_i^* \end{array} \right\} =$$

$$\dfrac{\begin{array}{c} \beta'^*, \mathcal{T}^*((q,\delta), \tau^*), \\ min\{\tau^*, \beta'^*\} = \beta^* \end{array}}{(q,\delta)\ R_\beta^*\ (\sigma\pi, \beta^*)}.$$

where $\sigma\pi$ are the substitutions that generate $\beta_\bullet$ and β^* and the ψ_i are non-empty conjunctions.

$$\dfrac{\begin{array}{c} (\neg g, \delta), g\ R_\beta^*\ (yes, \beta'), \beta' < 1, \delta\ R_{\beta_\bullet}\ (\sigma, \beta''), \\ \mathcal{T}_\bullet((\neg g, \delta), \tau_\bullet), max\{\tau_\bullet, (1 - \beta') + \beta'' - 1\} = \beta_\bullet \end{array}}{(\neg g, \delta)\ R_{\beta_\bullet}\ (\sigma, \beta_\bullet)}$$

$$\dfrac{\begin{array}{c} (\neg g, \delta), g\ R_{\beta_\bullet}\ (yes, \beta'), \beta' < 1, \delta\ R_\beta^*\ (\sigma, \beta''), \\ \mathcal{T}^*((\neg g, \delta), \tau^*), min\{\tau^*, (1 - \beta'), \beta''\} = \beta^* \end{array}}{(\neg g, \delta)\ R_\beta^*\ (\sigma, \beta^*)}$$

3. $$\dfrac{(\neg g, \delta), gF_\beta^*, \delta\ R_{\beta_\bullet}\ (\sigma, \beta_\bullet)}{(\neg g, \delta)\ R_{\beta_\bullet}\ (\sigma, \beta_\bullet)}.$$
   $$\dfrac{(\neg g, \delta), gF_{\beta_\bullet}, \delta\ R_\beta^*\ (\sigma, \beta^*)}{(\neg g, \delta)\ R_\beta^*\ (\sigma, \beta^*)}.$$

4. $$\dfrac{(q, \delta), \exists[p \leftarrow \psi]: \exists mgu(q, p)}{(q, \delta)\ F_{\beta_\bullet}}.$$
   $$\dfrac{(q, \delta), \exists[p \leftarrow \psi]: \exists mgu(q, p)}{(q, \delta)\ F_\beta^*}.$$
   $$\dfrac{(q, \delta), \forall[p_i \leftarrow \psi_i]: \exists \sigma = mgu(q, p_i) \Rightarrow (\psi_i, \delta)\sigma\ F_{\beta_\bullet}}{(q, \delta)\ F_{\beta_\bullet}}.$$
   $$\dfrac{(q, \delta), \forall[p_i \leftarrow \psi_i]: \exists \sigma = mgu(q, p_i) \Rightarrow (\psi_i, \delta)\sigma\ F_\beta^*}{(q, \delta)\ F_\beta^*}.$$

5. $$\dfrac{(\neg g, \delta), g\ R_\beta^*\ (yes, 1)}{(\neg g, \delta)\ F_{\beta_\bullet}}.$$
   $$\dfrac{(\neg g, \delta), g\ R_{\beta_\bullet}\ (yes, 1)}{(\neg g, \delta)\ F_\beta^*}.$$

Fig. 6. β-SLDNF

6.2 Dealing with conditional beliefs

We may want to constrain our queries to a specific set of possible worlds in which a statement is believed to be (to some extent) true. In other words, we may be interested in measuring *conditional beliefs* on queries.

If we had the values for the inner and outer measures $\mathcal{B}_*(\psi)$ and $\mathcal{B}^*(\psi)$, we could evaluate conditional beliefs by using the expressions given in (Fagin and Halpern 1989b):

$$\mathcal{B}_*(\psi_1|\psi_2) = \begin{cases} \frac{\mathcal{B}_*(\psi_1,\psi_2)}{\mathcal{B}_*(\psi_1,\psi_2)+\mathcal{B}^*(\neg\psi_1,\psi_2)}, & \mathcal{B}_*(\psi_1,\psi_2)+\mathcal{B}^*(\neg\psi_1,\psi_2)\neq 0 \\ 0, & \mathcal{B}_*(\psi_1,\psi_2)+\mathcal{B}^*(\neg\psi_1,\psi_2)=0 \end{cases}$$

$$\mathcal{B}^*(\psi_1|\psi_2) = \begin{cases} \frac{\mathcal{B}^*(\psi_1,\psi_2)}{\mathcal{B}^*(\psi_1,\psi_2)+\mathcal{B}_*(\neg\psi_1,\psi_2)}, & \mathcal{B}^*(\psi_1,\psi_2)+\mathcal{B}_*(\neg\psi_1,\psi_2)\neq 0 \\ 0, & \mathcal{B}^*(\psi_1,\psi_2)+\mathcal{B}_*(\neg\psi_1,\psi_2)=0 \end{cases}$$

However, we can only access the values $\hat{\mathcal{B}}_*(\psi) \leq \mathcal{B}_*(\psi)$ and $\hat{\mathcal{B}}^*(\psi) \geq \mathcal{B}^*(\psi)$ for a clause ψ. Since we have that (Fagin and Halpern 1989b):

$$[\mathcal{B}_*(\psi_1|\psi_2), \mathcal{B}^*(\psi_1|\psi_2)] \subseteq [\frac{\mathcal{B}_*(\psi_1,\psi_2)}{\mathcal{B}^*(\psi_2)}, \frac{\mathcal{B}^*(\psi_1,\psi_2)}{\mathcal{B}_*(\psi_2)}]$$

We immediately have that:

$$[\mathcal{B}_*(\psi_1|\psi_2), \mathcal{B}^*(\psi_1|\psi_2)] \subseteq [\frac{\hat{\mathcal{B}}_*(\psi_1,\psi_2)}{\hat{\mathcal{B}}^*(\psi_2)}, \frac{\hat{\mathcal{B}}^*(\psi_1,\psi_2)}{\hat{\mathcal{B}}_*(\psi_2)}]$$

Hence, we adopt these expressions as approximations for the lower and upper bounds for conditional degrees of belief:

- $\hat{\mathcal{B}}_*(\psi_1|\psi_2) = \frac{\hat{\mathcal{B}}_*(\psi_1,\psi_2)}{\hat{\mathcal{B}}^*(\psi_2)}$,
- $\hat{\mathcal{B}}^*(\psi_1|\psi_2) = min\{1, \frac{\hat{\mathcal{B}}^*(\psi_1,\psi_2)}{\hat{\mathcal{B}}_*(\psi_2)}\}$

7 Summary and Discussion

There has been a lot of debate on which formalism to measure uncertainty is the most general, and many researchers have recently defended the view that there is not a most general formalism but that different formalisms are better to measure different facets of uncertainty.

In the present work we adopted the latter view. We also avoided the simplification that *a single* facet of uncertainty should be selected at the end, therefore accepting that multiple measures could have to be considered within a single representation language.

Assuming this point of view, we explored the feasibility of performing automated reasoning about a domain containing more than one facet of uncertainty by (i) selecting three of these facets and their corresponding measures; (ii) incorporating them in a resolution-based, first-order, clausal theorem prover; and (iii) implementing this theorem prover as a PROLOG meta-interpreter.

An important aspect of any knowledge representation schema for automated reasoning is to have a clearly and rigorously specified semantics for its expressions and operations, so we were careful about providing a model theory for our language and guaranteeing the soundness of its inference procedures.

The main expected contribution of this work was the evidential proof that multiple measures of uncertainty are useful in knowledge representation and inference, and that they can (and should) be treated conjointly within a single representation language. Nevertheless, we believe that the language which was constructed and implemented to constitute this proof presents interest in itself as the prototype of a language to implement knowledge-based systems about domains pervaded with uncertainty. With this in mind, we explored some possibilities to improve its computational efficiency in time, which presented some positive results.

One aspect of our language is the variable coarseness of the results it produces. The language is a proper extension of several simpler theorem provers (e.g., Lee's language (1972), Halpern's logic (1990), the logic Lp (Bacchus 1988, Bacchus 1990a), Nilsson's logic (1986), and when "projected" onto one of these languages it produces results at least as precise as those (i.e., if the result is an interval, it is going to be at least as tight as the one produced by the simpler language). In those more complex cases in which the extensions are needed, however, the uncertainty intervals generated by our language grow rapidly in width. It is a topic for further research whether we can specify particular classes of problems with special structural properties such that more precise results (i.e. tighter intervals) can be obtained.

Another limitation of the language is the presupposition of a single source of information for a program (i.e., a single agent to assign belief and truth-degrees to expressions), and the consideration of only those problems which can be treated monotonically. It is still an open question whether a richer language, capable of treating non-monotonicities and multiple agents (which can be independent, partially dependent or totally dependent), can be constructed in such a way that the language has a clear declarative semantics and is computationally tractable.

Acknowledgements

The authors profited from comments and suggestions from many researchers along the development of this work. We would like to thank especially Fahiem Bacchus, Alan Bundy, Didier Dubois, John Fox, Jérôme Lang, Jeff Paris, Henri Prade, Sandra Sandri and V.S. Subrahmanian. The first author was sponsored by CNPq - Conselho Nacional de Desenvolvimento Científico e Tecnológico (Brazil) - grant nr. 203004/89.2. The final production and presentation of this paper was partially supported by FAPESP - Fundação de Amparo à Pesquisa no Estado de São Paulo (Brazil) - grant nr 92/4323-0.

References

Abadi, M., and J.Y. Halpern: *Decidability and Expressiveness for First-Order Logics of Probability.* IBM Research Report **RJ 7220**, 1989.

Apt, K.F.: *Introduction to Logic Programming.* Centre for Mathematics and Computer Science Report **CSR 8741**, 1987.

Bacchus, F.: *Representing and Reasoning with Probabilistic Knowledge.* University of Alberta, 1988.

Bacchus, F.: "Lp, a Logic for Representing and Reasoning with Statistical Knowledge," in: *Computational Intelligence* 6 (1990a) 209-231.

Bacchus, F.: *Representing and Reasoning with Probabilistic Knowledge.* MIT Press, 1990b.

Bundy, A.: "Incidence Calculus: a Mechanism for Probabilistic Reasoning," in: *Journal of Automated Reasoning* 1 (1985) 263-284.

Corrêa da Silva, F.S.: *Automated Reasoning with Uncertainties.* University of Edinburgh, Department of Artificial Intelligence, 1992.

Corrêa da Silva, F.S., and A. Bundy: *On Some Equivalence Relations Between Incidence Calculus and Dempster-Shafer Theory of Evidence.* 6th Conference on Uncertainty in Artificial Intelligence, 1990.

Corrêa da Silva, F.S., and A. Bundy: *A Rational Reconstruction of Incidence Calculus.* University of Edinburgh, Department of Artificial: Intelligence Report **517**, 1991.

Dubois, D., and H. Prade: "An Introduction to Possibilistic and Fuzzy Logics," in: P. Smets et al. (eds.), *Non-standard Logics for Automated Reasoning* Academic Press, 1988.

Dubois, D., and H. Prade: "Fuzzy Sets, Probability and Measurement," in: *European Journal of Operational Research* 40 (1989) 135-154.

Dudley, R.M.: *Real Analysis and Probability.* Wadsworth & Brooks/Cole, 1989.

Fagin, R., and J.Y. Halpern: *Uncertainty, Belief, and Probability.* IBM Research Report **RJ 6191**, 1989a.

Fagin, R., and J.Y. Halpern: *A New Approach to Updating Beliefs.* IBM Research Report **RJ 7222**, 1989.

Fitting, M.: "Logic Programming on a Topological Bilattice," in: *Fundamenta Informaticae* **XI** (1988) 209-218.

Fitting, M.: "Bilattices in Logic Programming," in: *Proceedings of the 20th International Symposium on Multiple-valued Logic*, 1990.

Halpern, J.Y.: "An Analysis of First-Order Logics of Probability," in: *Artificial Intelligence* **46** (1990) 311-350.

Halpern, J. Y., and R. Fagin: *Two Views of Belief: Belief as Generalised Probability and Belief as Evidence.* IBM Research Report **RJ 7221**, 1989.

Hinde, C.J.: "Fuzzy Prolog," in: *International Journal of Man-Machine Studies* **24** (1986) 569-595.

Ishizuka, M., and K. Kanai: "Prolog-ELF Incorporating Fuzzy Logic." in: *IJCAI'85 - Proceedings of the 9th International Joint Conference on Artificial Intelligence*, 1985.

Kifer, M., and V.S. Subrahmanian: "Theory of Generalized Annotated Logic Programs and Its Applications," in: *Journal of Logic Programming* **12**, 1991.

Klement, E.P.: "Construction of Fuzzy σ-algebras Using Triangular Norms," in: *Journal of Mathematical Analysis and Applications* **85** (1982) 543-565.

Kunen, K.: "Signed Data Dependencies in Logic Programs," in: *Journal of Logic Programming* **7** (1989) 231-245.

Lee, R.C.T.: "Fuzzy Logic and the Resolution Principle," in: *Journal of the ACM* **19** (1972) 109-119.

Mendelson, E.: *Introduction to Mathematical Logic* (3rd. ed). Wadsworth & Brooks/Cole, 1987.

Nilsson, N.J.: "Probabilistic Logic," in: *Artificial Intelligence* **28** (1986) 71-87.

Ng, R., and V.S. Subrahmanian: "Probabilistic Logic Programming," in: *Information and Computation* **101**, 1992.

Orci, I.P.: "Programming in Possibilistic Logic," in: *International Journal of Expert Systems* **2** (1989) 79-96.

Piasecki, K.: "Fuzzy p-Measures and their Application in Decision Making," in: J. Kacprzyk and M. Fedrizzi (eds.), *Combining Fuzzy Imprecision with Probabilistic Uncertainty in Decision Making.* Berlin heidelberg: Springer Verlag, 1988.

Ruspini, E.H.: "On the Semantics of Fuzzy Logic," in: *SRI International* **475**, 1989.

Saffiotti, A.: "An AI View of the Treatment of Uncertainty," in: *The Knowledge Engineering Review* **2** (1987) 75-97.

Shafer, G.: *A Mathematical Theory of Evidence.* Princeton University Press, 1976.

Shapiro, E.Y.: "Logic Programming with Uncertainties - a Tool for Implementing Rule-based Systems," in: *IJCAI'83 - Proceedings of the 8th International Joint Conference on Artificial Intelligence*, 1983.

Shoenfield, J.R.: *Mathematical Logic.* Addison-Wesley, 1967.

Smets, P.: "Probability of a Fuzzy Event: An Axiomatic Approach," in: *Fuzzy Sets and Systems* **7** (1982) 153-164.

Turi, D.: *Logic Programs with Negation: Classes, Models, Interpreters.* Centre for Mathematics and Computer Science Report **CSR 8943**, 1989.

Turksen, I.B.: "Stochastic Fuzzy Sets: a Survey," in: J. Kacprzyk and M. Fedrizzi (eds.), *Combining Fuzzy Imprecision with Probabilistic Uncertainty in Decision Making.* Berlin, Heidelberg: Springer Verlag, 1988.

Van Emden, M.H.: "Quantitative Deduction and its Fixpoint Theory," in: *Journal of Logic Programming* **1** (1986) 37-53.

An Axiomatic Approach to Systems of Prior Distributions in Inexact Reasoning

Jonathan Lawry* and George M. Wilmers

Department of Mathematics, University of Manchester
Manchester M13 9PL, U.K.
E-mail: mbbgpla@hpa.mcc.ak.uk/george@ma.man.ac.uk

Abstract. We describe an axiomatic approach to the *a priori* choice of hierarchies of second order probability distributions within the context of inexact reasoning. In this manner we give an epistemological characterisation of a certain hierarchy of symmetric Dirichlet priors up to a parameter.

1 Motivation

One of the fundamental problems in the development of expert systems is how to make decisions based on uncertain or inexact knowledge. For example, an expert system for medical diagnosis and prognosis might consist of a knowledge base as follows :

- 80% of patients with symptoms S have disease D
- Between 2% and 5% of patients are of blood group B and have disease D
- Patients of blood group B with symptom S are more likely than not to have disease D

A wide range of such inexact knowledge bases occurring in practice may reasonably be represented by linear equality or inequality constraints on the subjective belief function of the expert from whom the information was collected. This assumes that each individual has a belief function which allocates a numerical value, usually normalised so that it lies between zero and one, to each proposition. For the following we shall consider that all belief functions are probability functions. Hence, the above example could be formulated as:

$$
\begin{aligned}
E(S \wedge D) &= 0.8 \times E(S) \\
0.02 \le E(B \wedge D) &\le 0.05 \\
E(S \wedge B \wedge D) &> 0.5 \times E(S \wedge B)
\end{aligned}
$$

for some probability function E.

We consider only propositional languages with binary connectives and negation where the literals are the fundamental observable phenomena relating to the domain of discussion. Clearly then, in most cases the knowledge base will

---

* Supported by a Science and Engineering Research Council Award

not provide sufficient information to uniquely determine a probability value for every sentence of such a language. Therefore, to select a single probability function defined on the whole language, we must make *a priori* assumptions. These take the form of an *inference process*, N, which for a given knowledge base C, defines for every sentence θ of the language a corresponding real value $N_C(\theta)$ consistent with C. This inference process may then be incorporated as part of the inference engine of the expert system.

Given the uncertainty regarding the probability function E, we feel that it is natural to define a second order distribution on belief functions (Paris et al. 1991, Paris et al., to appear). This accords with the feeling that human experts do not always precisely define a subjective belief function but rather have some general intuition regarding their beliefs relative to a knowledge base. Thus, an inference process to select a particular belief function need only be used when a precise decision is required. We aim to develop epistemologically based systems of axioms identifying particular classes of second order distributions. Such classes of priors are naturally applicable to many areas of inexact reasoning; in particular, any such class of second order distributions naturally defines an inference process the properties of which can be investigated.

2 Mathematical Formulation

We use the formulation of inference process developed by Paris and Vencovska in a number of papers including (Paris and Vencovska 1991). Let $L^{(n)}$ be a language of the propositional calculus consisting of the propositional variables $p_1, \ldots, p_n$ with connectives $\wedge, \vee$ and $\neg$. The sentences of $L^{(n)}$ are denoted by $SL^{(n)}$. Now a probability function on $SL^{(n)}$ is a function E from $SL^{(n)}$ into $[0, 1]$ which satisfies the following axioms:

(i) $\models \theta \Longrightarrow E(\theta) = 1$

(ii)   if $\models \neg(\theta \wedge \varphi)$ then $E(\theta \vee \varphi) = E(\theta) + E(\varphi)$

which together imply:

(iii)   if $\models \theta \leftrightarrow \varphi$ then $E(\theta) = E(\varphi)$

The condition (iii) means that we can define E on the Lindenbaum algebra, $\mathcal{L}^{(n)}$, of equivalence classes for the equivalence relation of logical equivalence (i.e., for $\varphi, \theta \in SL^{(n)}$ the algebra consists of equivalence classes $\overline{\theta}$ such that $\varphi \in \overline{\theta}$ iff $\varphi \leftrightarrow \theta$). Now by the disjunctive normal form theorem, every sentence of $L^{(n)}$ is equivalent to a disjunction of sentences of the form $\bigwedge_i^n p_i^{\epsilon i}$ where $\epsilon_i \in \{0, 1\}$ and $p^1 = p, p^0 = \neg p$. The equivalence classes containing these sentences form the atoms of $\mathcal{L}^{(n)}$ relative to the natural ordering on boolean algebras and hence we refer to them as atoms. Clearly for a language of size n there are 2^n atoms which we enumerate by

$$\alpha_j^{(n)} = \bigwedge_{i=1}^{n} p_i^{\epsilon_i^j}$$

where

$$j = 2^n - \sum_{i=1}^{n} \epsilon_i^j 2^{n-i}$$

Since the $\bar{\alpha}_i^{(n)}$ are disjoint and $\bigvee_i^{2^n} \bar{\alpha}_i^{(n)} = \top$ it follows that a probability function E on $\mathcal{L}^{(n)}$ is uniquely defined by its values on the $\alpha_i^{(n)}$ for $i = 1, \ldots, 2^n$. Therefore our knowledge base can be expressed in terms of linear constraints on $E(\bar{\alpha}_i^{(n)})$ for $i = 1, \ldots, 2^n$. Further, if we define

$$V^{(n)} = \{\mathbf{x} \in [0, 1]^{2^n} | \sum_{i=1}^{2^n} x_i = 1\} \subset [0, 1]^{2^n}$$

then there is a one to one correspondence from points in $V^{(n)}$ onto probability functions on $\mathcal{L}^{(n)}$. This is to say that for all $\mathbf{x}$ in $V^{(n)}$, $\mathbf{x}$ defines a probabability function on $\mathcal{L}^{(n)}$ such that $E(\bar{\alpha}_i^{(n)}) = x_i$ and for all probability functions E with $E(\bar{\alpha}_i^{(n)}) = x_i$, $\mathbf{x}$ is in $V^{(n)}$. To avoid clumsy notation we associate sentences of $L^{(n)}$ with their equivalence classes in $\mathcal{L}^{(n)}$ from now on.

Let $\mathcal{V}^{(n)}$ be the σ-algebra of Borel sets generated by the intervals of $V^{(n)}$. Then any *a priori* choice of a second order distribution corresponds to a choice of probability function, $Prob^{(n)}$, on $\mathcal{V}^{(n)}$. For the scope of this paper we assume that $Prob^{(n)}$ has a density function $f^{(n)}$ such that for all R in $\mathcal{V}^{(n)}$

$$Prob^{(n)}(R) = \int_R f^{(n)}(\mathbf{x}) dV^{(n)}$$

where integration is carried out with respect to the uniform measure on $\mathcal{V}^{(n)}$.

Definition 1 *Let $\mathcal{F}$ be the class of all density functions which are continuous on the interior of $V^{(n)}$.*

Henceforth, we make a smoothness assumption in this paper by restricting our attention to densities in $\mathcal{F}$.

If C is a knowledge base of linear constraints, then C corresponds to a convex subset of $V^{(n)}$ denoted also by C. Therefore, any $f^{(n)}$ in $\mathcal{F}$ which is integrable on C naturally defines a density function on C given by

$$\frac{f^{(n)}(\mathbf{x})}{\int_C f^{(n)}(\mathbf{x}) dV^{(n)}}$$

where by notational convention integration is carried out in relative dimension if C is a lower dimensional region of $V^{(n)}$. If, in this case, the integral of $f^{(n)}$ on C does not exist, then for the scope of this paper we leave the conditional density function undefined.

With the same convention we define for any R in $\mathcal{V}^{(n)}$

$$Prob^{(n)}(R|C) = \frac{\int_{R \cap C} f^{(n)}(\mathbf{x}) dV^{(n)}}{\int_C f^{(n)}(\mathbf{x}) dV^{(n)}}$$

In this context a rather natural inference process is the **Center of Mass inference process** $(CM^{f^{(n)}})$, so called because it selects the center of mass of a convex region C relative to a prior density $f^{(n)}$. More formally , if

$$CM_C^{f^{(n)}}(\alpha_i^{(n)}) = \frac{\int_C x_i f^{(n)}(\mathbf{x}) dV^{(n)}}{\int_C f^{(n)}(\mathbf{x}) dV^{(n)}} = y_i$$

for $i = 1, \ldots, 2^n$, then $\mathbf{y}$ is the center of mass of the region C calculated according to the density $f^{(n)}$. Since C is convex, $\mathbf{y}$ is a point in C, and hence the inference process selects a probability function E consistent with the knowledge base C. Note, however, that if the dimension of C is less than $2^n - 1$ then $f^{(n)}$ may not be integrable on C in which case $CM_C^{f^{(n)}}(\alpha_i^{(n)})$ is undefined for $i = 1, \ldots, 2^n$. The inference process $CM^{f^{(n)}}$ for the case when $f^{(n)}$ is the uniform measure, has been investigated by several authors (e.g. Paris and Vencovska (1991, to appear)).

Up to now we have considered the language $L^{(n)}$ as fixed. However, if we allow n to vary by adding propositional variables so that $L^{(n)} = \{p_1, \ldots, p_n\}$, and then for each value of $n \geq 1$ define a probability measure $Prob^{(n)}$ with density $f^{(n)}$, we obtain a hierarchy of prior densities denoted $\{f^{(n)}\}$. In the next section we propose an axiomatic framework for the *a priori* choice of a hierarchy of densities, and then go on to give some results related to this sytem of axioms.

3 Axiomatic Framework

The problem of choosing second order densities has parallels with the old statistical problem of defining priors for use in Bayesian inference. It has often been pointed out that for some types of Bayesian reasoning the choice of prior is relatively unimportant assuming large sample sizes. Furthermore, Bayesian statisticians have, in spite of the arguments put forward by (Johnson 1932) and (Carnap 1952), tended to adopt a rather ad hoc approach to selecting priors. In the present context of inexact reasoning, however, the choice of priors is of fundamental importance and hence we feel it is necessary to develop an approach based on epistemological considerations.

The traditional response to the problem, attributed to Laplace in the case $n = 1$ (see (Zabel 1965) for discussion), is that in the absence of any other information the principle of insufficient reason forces us to choose $f^{(n)}$ to be the uniform distribution (i.e $Prob^{(n)}(R)$ proportional to the volume of R for $R \in \mathcal{V}^{(n)}$). However, we shall demonstrate, at least for the more general n dimensional problem, that this is unsatisfactory.

The epistemological conditions that we propose can be separated into two categories: **global conditions** which describe relationships between different levels of the hierarchy, and **local conditions** which are defined for fixed n. The first axiom, marginality, is a global condition.

A1 (Marginality) For all $n > 1$ if $\alpha_i^{(n-1)}$ for $i = 1, \ldots, 2^{n-1}$ denote the atoms of $L^{(n-1)}$ then

$$Prob^{(n)}(\bigwedge_{i=1}^{2^{n-1}} E(\alpha_i^{(n-1)}) < t_i) = Prob^{(n-1)}(\bigwedge_{i=1}^{2^{n-1}} E(\alpha_i^{(n-1)}) < t_i)$$

Expressed in terms of densities: let $\{f^{(n)}\}$ be a hierarchy of prior densities and let

$$Y_i = X_{2i-1} + X_{2i} \; for \; i = 1, \ldots, 2^{n-1}$$

If $h^{(n)}(\mathbf{y})$ is the marginal density function of $\mathbf{Y}$ induced by $f^{(n)}$, then $\{f^{(n)}\}$ is said to satisfy **marginality** if

$$\forall \, n > 1 \; h^{(n)}(\mathbf{y}) = f^{(n-1)}(\mathbf{y}).$$

Note that in our enumeration the atoms of $L^{(n-1)}$ are related to those of $L^{(n)}$ by

$$\alpha_{2j}^{(n)} \vee \alpha_{2j-1}^{(n)} = \alpha_j^{(n-1)} \; where \; \alpha_{2j-1}^{(n)} = \alpha_j^{(n-1)} \wedge p_n \; and \; \alpha_{2j}^{(n)} = \alpha_j^{(n-1)} \wedge \neg p_n$$

This axiom is motivated by the consideration that adding another propositional variable p_n to the language $L^{(n-1)}$ should not effect our knowledge of sentences that do not contain p_n. The axiom of marginality is sufficient to guarantee this in the presence of an axiom which allows the permutation of variables (see axiom A2). Marginality is strongly related to the following condition on inference processes.

Definition 2 *Suppose* $N(L^{(n)})$ *is a family of inference processes dependent on the language* $L^{(n)}$. *This family is said to be* **language invariant** *if for* $L^{(n)} \subseteq L^{(n+m)}$ *and* C *a set of constraints on* $L^{(n)}$

$$N_C(L^{(n)})(\theta) = N_C(L^{(n+m)})(\theta) \; \forall \theta \in SL^{(n)}$$

Theorem 3 *If* $f^{(n)} \in \mathcal{F}$ *and* $f^{(n)}(\mathbf{x}) > 0$ *for all* $\mathbf{x}$ *in the interior of* $V^{(n)}$, *then* $CM^{f^{(n)}}$ *satisfies language invariance, iff* $\{f^{(n)}\}$ *satisfies marginality.*

We can now see that the assumption of the uniform measure is unsatisfactory in that it is inconsistent with marginality. In other words, if $Prob^{(n)}$ is defined to be the uniform measure at each level n then the hierarchy of priors does not satisfy marginality. In addition, we observe that the uniform density is a special case of the symmetric Dirichlet system of densities given by

$$d(\lambda, 2^n) = \Gamma(\lambda)[\Gamma(\frac{\lambda}{2^n})^{-2^n}] \prod_{i=1}^{2^n} x_i^{\frac{\lambda}{2^n} - 1}$$

where λ is a parameter on $(0, \infty)$ corresponding to the uniform distribution when $\lambda = 2^n$. This system of priors is frequently used in Bayesian reasoning and has been justified in this context by (Johnson 1932) and (Carnap 1952). The inconsistency of the uniform measure with marginality is clarified by the next result.

Theorem 4 *If $\{d(\lambda, 2^n)\}$ is a hierarchy of symmetric Dirichlet priors then $\{d(\lambda, 2^n)\}$ satisfies marginality iff $\lambda > 0$ is a constant independent of n.*

Our second axiom is the local symmetry condition that the density $f^{(n)}$ should be invariant under permutations of the literals of the language.

A2 (Weak Renaming) Let $\sigma : \mathcal{L}^{(n)} \longmapsto \mathcal{L}^{(n)}$ be an automorphism on the Lindenbaum algebra such that for all propositional variables p_i

$$\sigma(p_i) = p_j^\epsilon \ where \ \epsilon \in \{0, 1\}$$

For each such σ there is a corresponding permutation of each $\mathbf{x} \in V^{(n)}$ such that

$$\sigma(x_i) = x_j \iff \sigma(\alpha_i^{(n)}) = \alpha_j^{(n)}$$

Then $f^{(n)}$ satisfies **weak renaming** if

$$\forall \mathbf{x} \in V^{(n)} \ f^{(n)}(\sigma(\mathbf{x})) = f^{(n)}(\mathbf{x})$$

This axiom seems intuitively justified because simply renaming the literals should not alter the *a priori* information content of the system.

A3 (Non Nullity)

$$Prob^{(n)}(E(\alpha_i^{(n)}) \in I) > 0 \ for \ all \ non \ empty \ open \ intervals \ I \subseteq [0, 1].$$

In terms of densities this means that for all such intervals $I \subseteq [0, 1]$ $\exists \mathbf{x} \in V^{(n)}$ such that $x_i \in I$ and $f^{(n)}(\mathbf{x}) > 0$.

A problem that seems to occur regularly in relation to knowledge bases for expert systems is as follows. Suppose we have total knowledge regarding specific atoms of $SL^{(n-1)}$; this is to say our knowledge base consists of constraints of the form $E(\alpha_i^{(n-1)}) = t$. If a new propositional variable p_n is then added to the language, and we have no prior information either of the weights to be allocated to the new atoms $p_n \wedge \alpha_i^{(n-1)}$ and $\neg p_n \wedge \alpha_i^{(n-1)}$, or what the relationship is between these and the other atoms of $L^{(n)}$, then we use the final two axioms.

A4 (Weak Atomic Independence) For $n \geq 2$ the distribution $Prob^{(n)}$ satisfies weak atomic independence if

$$Prob^{(n)}(E(p_n | \alpha_i^{(n-1)}) < x \ |E(\alpha_i^{(n-1)}) = t \ )$$

$$= Prob^{(n)}(E(p_n | \alpha_i^{(n-1)}) < x \ | \bigwedge_{j \neq i}(E(\alpha_{2j}^{(n)}) = s_{2j} \wedge E(\alpha_{2j-1}^{(n)}) = s_{2j-1} \ ))$$

for all $s_{2j}, s_{2j-1}, t, x \in [0, 1]$ where $j = 1, \ldots, 2^{n-1}$ subject to $\sum_{j \neq i} s_{2j} + s_{2j-1} = 1 - t$.

Less formally, weak atomic independence states that for all $j \in \{1, \ldots, 2^{n-1}\}$ $E(\alpha_{2i}^{(n)})$ and $E(\alpha_{2i-1}^{(n)})$ are independent of the probability of the other atoms of $L^{(n)}$ up to consistency with the constraint $E(\alpha_i^{(n-1)}) = t$.

A5 (Relative Ignorance) The distribution $Prob^{(n)}$ satisfies relative ignorance if for $x, t \in [0, 1], n \geq 2$

$$Prob^{(n)}(E(p_n | \alpha_i^{(n-1)}) < x | E(\alpha_i^{(n-1)}) = t)$$

is independent of t.

In other words the probability of p_n occurring, given that $\alpha_i^{(n-1)}$ has occurred, is independent of $E(\alpha_i^{(n-1)})$. The idea here is that since the constraint $E(\alpha_i^{(n-1)}) = t$ provides no information about p_n, the relative distribution of weight between the atoms $p_n \wedge \alpha_i^{(n-1)}$ and $\neg p_n \wedge \alpha_i^{(n-1)}$ should be independent of t. The next section gives a number of results related to this axiom system.

4 Multiplicativity and Dirichlet Priors

The prior densities considered in this section have the form $f^{(n)} = \prod_{i=1}^{2^n} g^{(n)}(x_i)$ where $g^{(n)} : [0, 1] \to \mathbf{R}^+$ and we refer to such priors as **multiplicative**. A good example are symmetric Dirichlet priors described earlier. The next result illustrates the connection between multiplicative priors and weak atomic independence. All results in this section concern priors in $\mathcal{F}$ which satisfy weak renaming and non nullity.

Theorem 5 *Let $\{f^n\}$ be a hierarchy of prior densities satisfying marginality. For all $n \geq 1$ let $f^{(n)}$ be a density in $\mathcal{F}$ which satisfies weak renaming and non nullity. Then the following are equivalent*

(i) $\forall n \geq 2$ $f^{(n)}$ *satisfies weak atomic independence*
(ii) $\forall n \geq 1$ $f^{(n)}$ *is multiplicative*

Thus assuming weak atomic independence we are able to further restrict our attention to multiplicative priors. In fact, we now limit ourselves to the following class of smooth multiplicative priors.

Definition 6 *Let $\mathcal{F}_\Pi$ be the class of multiplicative priors of the form $f^{(n)}(\mathbf{x}) = \prod_{i=1}^{2^n} g(x_i)$ where $g : (0, 1) \to \mathbf{R}^+$ is twice differentiable on $(0, 1)$.*

We now give a result demonstrating the equivalence, in this context, between accepting the axiom of relative ignorance and restricting attention to symmetric Dirichlet priors.

Theorem 7 *If $f^{(n)}$ is in $\mathcal{F}_{\Pi}$ and satisfies non nullity, then $f^{(n)}$ satisfies relative ignorance iff $f^{(n)}$ is a symmetric Dirichlet prior.*

The following corollary shows that for the class of densities $\mathcal{F}_{\Pi}$, the axioms A1, A2, A3 and A5 characterise the hierarchy of symmetric Dirichlet priors up to a parameter λ independent of n.

Corollary 8 *If $\{f^{(n)}\}$ is a hierarchy of priors satisfying marginality such that for all $n \geq 1$ $f^{(n)}$ is in $\mathcal{F}_{\Pi}$ and satisfies non nullity, weak renaming, and relative ignorance, then $f^{(n)}(\mathbf{x}) = d(\lambda, 2^n)(\mathbf{x})$ where $\lambda > 0$ is a constant independent of n.*

5 Comments

From a theorem of De Finetti it follows that Carnap's continuum for inductive inference on sequences of trials with n possible outcomes is equivalent to Bayesian inference assuming a symmetric Dirichlet prior (see (Carnap 1952) or (Zabel 1965)). In view of this and the above results an open question worthy of consideration is what relationship exists between axioms A1 to A5 and the logical conditions of Carnap? For instance, one symmetry principle of Carnap's states that the probability of a particular propositional variable, p, occurring at the next trial is dependent only on the total number of trials and the number of occurrences of p in previous trials. There has been much criticism of this principle on the basis that it fails to take into account information on frequencies of frequencies which can be used to estimate the probabilities of literals (Good 1965). In the light of this it would be interesting to know if analogous criticisms can be made regarding A4 and A5. In addition, given the failure of Carnap's programme to justify the choice of a value for λ, it would be interesting to know if there are natural axioms for systems of priors which together with axioms A1 to A5 would choose a particular λ.

It is important to note that we are not claiming the above axioms are natural laws justifying Dirichlet priors, neither is it our intention to propound any one system of axioms. Rather we aim to demonstrate the relationship between particular hierarchies of priors and epistemological conditions. Indeed, we are investigating a number of other axioms, some of which are inconsistent with A1 to A5. For example:

Epistemic Independence

$$Prob^{(n)}(E(p_i) < t_i \mid \bigwedge_{j \neq i} E(p_j) < t_j) = Prob^{(n)}(E(p_i) < t_i)$$

where $j = 1, \ldots, n$

This condition states that $E(p_j)$ for $j \neq i$ provides no information about $E(p_i)$ and is based on the idea that since we have no knowledge of relationships between the propositional variables we assume them independent.

Hierarchical Independence For $n \geq 2$

$$Prob^{(n)}(E(p_n | \alpha_i^{(n-1)}) < x \; | E(\alpha_i^{(n-1)}) = t \;)$$

is independent of n.

This is motivated by the feeling that for constraints $E(\alpha_i^{(n-1)}) = t$ the degree of uncertainty regarding $E(p_n \wedge \alpha_i^{(n-1)})$ is the same for each level of the hierarchy.

It can be shown that both hierarchical and epistemic independence are inconsistent with A1 to A5 for densities in $\mathcal{F}_\Pi$. Hierarchical independence together with A2, A3 and A5 characterise the hierarchy of symmetric Dirichlet priors $\{d(\kappa 2^n, 2^n)\}$ where $\kappa > 0$ is a constant. Clearly this is inconsistent with the hierarchy characterised by A1, A2, A3 and A5 (corollary 8). Similarly we can show that any hierarchy of priors in $\mathcal{F}_\Pi$ satisfying epistemic independence and A1 to A3 is also inconsistent with the hierarchy of corollary 8.

References

Carnap, R.: *Continuum of Inductive Methods.* Chicago: University of Chicago Press, 1952.

Good, I.J.: *The Estimation of Probabilities: An Essay on Modern Bayesian Methods.* M.I.T Press, 1965.

Good, I.J.: *Good Thinking: The Foundations of Probability and its Applications.* University of Minnesota Press, 1983.

Johnson, W.E.: "Probability: The Deductive and Inductive Problems," in: *Mind*, Vol. XLI, No. 164 (1932) 27.

Paris, J., and A. Vencovska: "A Method for Updating Justifying Minimum Cross Entropy," in: *The Journal of Approximate Reasoning*, to appear.

Paris, J., and A. Vencovska: "Principles of Uncertain Reasoning," in: *Proceedings of the second International Colloquium on Cognitive Science*, San Sebastian, Spain, 1991.

Paris, J., A. Vencovska, and G.M. Wilmers: "A Note on Objective Inductive Inference," in: De Glas, M., and D, Gabbay (eds.). *Proceedings of the First World Conference on the Foundations of Artificial Intelligence* Paris: Association Francaise pour l'Intelligence Artificialle (1991) 407-412.

Paris, J., A. Vencovska, and G.M. Wilmers: "A Natural Prior Probability Distribution Derived From The Propositional Calculus," in: *The Annals of Pure and Applied Logic*, to appear.

Zabel, S.L.: "Symmetry and its Discontents," in: *Causation, Chance, and Credence*, Vol 1. M.I.T. Press (1965) 155-190.

Contradiction Removal Semantics
with Explicit Negation

Luís Moniz Pereira, José J. Alferes, Joaquim N. Aparício

CRIA, Uninova and DCS, U. Nova de Lisboa
2825 Monte da Caparica, Portugal
E-mail: {lmp/jja/jna}@fct.unl.pty

Abstract. Well Founded Semantics for logic programs extended with eXplicit negation ($WFSX$) is characterized by the fact that, in any model, whenever $\neg a$ (the explicit negation of a) holds, then $\sim a$ (the negation by default of a) also holds.

When explicit negation is used, contradictions may be present (e.g., a and $\neg a$ both hold for some a). We introduce a way the notion of removing some contradictions through identifying the set of models obtained by revising closed world assumptions. One such unique model is singled out as the contradiction removal semantics ($CRSX$). When contradictions do not arise, the contradiction removal semantics coincides with $WFSX$.

1 Introduction

We begin the paper by briefly reviewing the $WFSX$ semantics introduced in (Pereira and Alferes 1992) and by presenting some examples when $WFSX$ semantics is used. For a motivation regarding $WFSX$, see the introduction to (Pereira et al. 1994 – in this volume). Since some programs may have no semantics because they contain a contradiction and $WSFX$ does not directly deal with contradictions, we introduce the $CRSX$, a process of identifying negative literals such as sources of inconsistencies, and show how contradictions may be removed. For each alternative way of removing contradiction in program P, we construct a program P' such that $WFSX(P')$ is consistent. We then present some properties concerning the $CRSX$ semantics.

2 Language

Given a first order language $Lang$ (Przymusinska and Przymusinski 1990), an extended logic program is a set of rules of the form

$$H \leftarrow B_1, \ldots, B_n, \sim C_1, \ldots, \sim C_m \qquad m \geq 0, n \geq 0$$

where $H, B_1, \ldots, B_n, C_1, \ldots, C_m$ are classical literals. A (syntactically) classical literal (or explicit literal) is either an atom A or its explicit negation $\neg A$. We also use the symbol $\neg$ to denote complementary literals in the sense of explicit negation. Thus $\neg\neg A = A$. The symbol $\sim$ stands for negation by default[1]. $\sim L$ is called a default literal. Literals are either classical or default. A set of rules stands for all its ground instances w.r.t. *Lang*. When $n = m = 0$ we may simply write H instead of $H \leftarrow$.

As in (Przymusinska and Przymusinski 1990), we expand our language by adding to it the proposition **u** such that for every interpretation I, $I(\mathbf{u}) = 1/2$. By a non-negative program we mean a program whose premises are either classical literals or **u**. Given a program P we denote by $\mathcal{H}_P$ (or simply $\mathcal{H}$) its Herbrand base. If S is a set of literals $\{L_1, \ldots, L_n\}$, by $\sim S$ we mean the set $\{\sim L | L \in S\}$.

If S is a set of literals then we say S is *contradictory* (resp. *inconsistent*) iff there is classical literal L such that $\{L, \neg L\} \subseteq S$ (resp. $\{L, \sim L\} \subseteq S$). In this case we also say that S *is contradictory w.r.t. to L*, (resp. S *is inconsistent w.r.t. to L*). S is agnostic w.r.t. L iff neither $L \in S$ nor $\sim L \in S$.

3 WFSX overview

In this section we briefly review $WFSX$ semantics for logic programs extended with explicit negation. For full details the reader is referred to (Pereira and Alferes 1992).

$WFSX$ follows from WFS plus one basic "coherence" requirement:

$$\neg L \Rightarrow \sim L \tag{1}$$

i.e., (if L is explicitly false, L must be false) for any explicit literal L.

Example 1. Consider program $P = \{a \leftarrow \sim b, \quad b \leftarrow \sim a, \quad \neg a \leftarrow\}$.

If $\neg a$ were to be simply considered as a new atom symbol, say a', and WFS used to define the semantics of P (as suggested in (Prymusinski 1990)), the result would be $\{\neg a, \sim \neg b\}$, so that $\neg a$ is true and a is undefined. We insist that $\sim a$ should hold, and that a should not hold, because $\neg a$ does. Accordingly, the WFSX of P is $\{\neg a, b, \sim a, \sim \neg b\}$, since b follows from $\sim a$.

We begin by providing a definition of interpretation for programs with explicit negation which incorporates coherence from the start.

Definition 1 Interpretation. By an interpretation I of a language *Lang* we mean any set $T \cup \sim F$[2], where T and F are disjoint subsets of classical literals over the Herbrand base, and if $\neg L \in T$ then $L \in F$ (coherence)[3]. The set T contains all ground classical literals *true* in I, the set F contains all ground

---

[1] This designation has been used in the literature instead of the more operational *"negation as failure (to prove)"*. Another appropriate designation is *"implicit negation"*, in contradistinction to *"explicit negation"*.

[2] By $\sim\{a_1, \ldots, a_n\}$ we mean $\{\sim a_1, \ldots, \sim a_n\}$.

[3] For any literal L, if L is explicitly false L must be false. Note that the complementary condition "if $L \in T$ then $\neg L \in F$" is implicit.

classical literals *false* in I. The truth value of the remaining classical literals is *undefined* (The truth value of a default literal $\sim L$ is the 3-valued complement of L.)

We next extend with an additional rule the P modulo I transformation of (Przymusinska and Przymusinski 1990), itself an extension of the Gelfond-Lifschitz modulo transformation, to account for coherence.

Definition 2 P/I transformation. Let P be an extended logic program and let I be an interpretation. By P/I we mean a program obtained from P by performing the following three operations for every atom A :

- Remove from P all rules containing a default premise $L =\sim A$ such that $A \in I$.
- Remove from P all rules containing a non-default premise L (resp. $\neg L$) such that $\neg L \in I$ (resp. $L \in I$).
- Remove from all remaining rules of P their default premises $L =\sim A$ such that $\sim A \in I$.
- Replace all the remaining default premises by proposition $\mathbf{u}$[4].

The resulting program P/I is by definition non-negative, and it always has a unique $least(P/I)$, where $least(P/I)$ is:

Definition 3 Least-operator. We define $least(P)$, where P is a non-negative program, as the set of literals $T\cup \sim F$ obtained as follows:

- Let P' be the non-negative program obtained by replacing in P every negative classical literal $\neg L$ by a new atomic symbol, say $'\neg\_L'$.
- Let $T'\cup \sim F'$ be the least 3-valued model of P'.
- $T\cup \sim F$ is obtained from $T'\cup \sim F'$ by reversing the replacements above.

The least 3-valued model of a non-negative program can be defined as the least fixpoint of the following generalization of the van Emden–Kowalski least model operator Ψ for definite logic programs:

Definition 4 Ψ^* operator. Suppose that P is a non-negative program, I is an interpretation of P and A is a ground atom. Then $\Psi^*(I)$ is an interpretation defined as follows:

- $\Psi^*(I)(A) = 1$ iff there is a rule $A \leftarrow A_1, \ldots, A_n$ in P such that $I(A_i) = 1$ for all $i \leq n$.
- $\Psi^*(I)(A) = 0$ iff for every rule $A \leftarrow A_1, \ldots, A_n$ there is an $i \leq n$ such that $I(A_i) = 0$.
- $\Psi^*(I)(A) = 1/2$, otherwise.

---

[4] The special proposition u is *undefined* in all interpretations.

To avoid incoherence, a partial operator is defined that transforms any non-contradictory set of literals into an interpretation, whenever no contradiction[5] is present.

Definition 5 The *Coh* operator. Let $I = T \cup {\sim} F$ be a set of literals such that T is not contradictory. We define $Coh(I) = I \cup {\sim}\{\neg L \mid L \in T\}$.

Definition 6 The Φ operator. Let P be a logic program and I an interpretation, and let $J = least(P/I)$. If $Coh(J)$ exists we define $\Phi_P(I) = Coh(J)$. Otherwise $\Phi_P(I)$ is not defined.

Example 2. For the program of example 1 we have

$$P/\{\neg a, b, {\sim} a, {\sim}\neg b\} = \{b \leftarrow ; \neg a \leftarrow\},$$

$$least(P/\{\neg a, b, {\sim} a, {\sim}\neg b\}) = \{\neg a, b, {\sim} a\}$$

and:

$$Coh(\{\neg a, b, {\sim} a\}) = \{\neg a, b, {\sim} a, {\sim}\neg b\}.$$

Definition 7 WFS with explicit negation. An interpretation I of an extended logic program P is called an Extended Stable Model (XSM) of P iff $\Phi_P(I) = I$. The F-least Extended Stable Model is called the Well Founded Model. The semantics of P is determined by the set of all $XSMs$ of P.

Example 3. Let $P = \{a \leftarrow {\sim} b, {\sim} c; b \leftarrow {\sim} a; \neg c \leftarrow {\sim} d\}$. This program has a least model $M_1 = \{{\sim} d, \neg c, {\sim} c, {\sim}\neg a, {\sim}\neg b, {\sim}\neg d\}$ and two Extended Stabel Models $M_2 = M_1 \cup \{{\sim} a, b\}$ and $M_3 = M_1 \cup \{a, {\sim} b\}$. Considering model M_1 we have for $P/M_1 = \{a \leftarrow \mathbf{u}; b \leftarrow \mathbf{u}; c' \leftarrow\}$, and

$$least(P/M_1) = J = \{{\sim} d, c', {\sim} a', {\sim} b', {\sim} d', {\sim} c\} = \{{\sim} d, {\sim}\neg d, \neg c, {\sim} c, {\sim}\neg a, {\sim}\neg b\}.$$

Example 4. Let P be:

$$a \leftarrow {\sim} a \ (i)$$
$$b \leftarrow {\sim} a \ (ii)$$
$$\neg b \leftarrow \quad (iii)$$

After the transformation, program P' has a rule $b' \leftarrow$, and there is no way in proving ${\sim} b$ from rules (i) and (ii). And we have $least(P'/\{b', {\sim} b, {\sim} a'\}) = \{b', {\sim} a'\} = M$ which corresponds to the model $\{\neg b, {\sim}\neg a\}$ if the coherence principle is not applied. In our case we have $Coh(M) = \{\neg b, {\sim} b, {\sim}\neg a\}$ which is the intended result.

Definition 8 Contradictory program. An extended logic program P is contradictory iff it has no model, i.e., there exists no interpretation I such that $\Phi_P(I) = I$.

---

[5] We say a set of literals S is contradictory iff for some literal L, $L \in S$ and $\neg L \in S$.

4 Revising contradictory extended logic programs

Once we introduce explicit negation programs are liable to be contradictory:

Example 5. Consider program $P = \{a \leftarrow \; ; \neg a \leftarrow \sim b\}$. Since we have no rules for b, by CWA it is natural to accept $\sim b$ as true. By the second rule in P we have $\neg a$, leading to an inconsistency with the fact a. Thus no set containing $\sim b$ may be a model of P.

We argue that the CWA may not be held of atom b since it leads to a contradiction. We show below how to revise[6] this form of contradiction by making a suitable revision of the incorrect CWA on b. The semantics we introduce identifies $\{a, \sim \neg a\}$ as the intended meaning of P, where b is revised to undefined. Assuming b false leads to a contradiction; revising it to true instead of undefined would not minimize the revised interpretation.

4.1 Contradictory well founded model

In order to revise possible contradictions we need to identify those contradictory sets implied by applications of CWA. The main idea is to compute all consequences of the program, even those leading to contradictions, as well as those arising from contradictions. The following example provides an intuitive preview of what we intend to capture:

Example 6. Consider program P :

$$a \leftarrow \sim b \;\text{(i)} \qquad d \leftarrow a \quad \text{(iii)}$$
$$\neg a \leftarrow \sim c \;\text{(ii)} \qquad e \leftarrow \neg a \;\text{(iv)}$$

1. $\sim b$ and $\sim c$ hold since there are no rules for either b or c
2. $\neg a$ and a hold from 1 and rules (i) and (ii)
3. $\sim a$ and $\sim \neg a$ hold from 2 and inference rule (1)
4. d and e hold from 2 and rules (iii) and (iv)
5. $\sim d$ and $\sim e$ hold from 3 and rules (iii) and (iv), as they are the only rules for d and e
6. $\sim \neg d$ and $\sim \neg e$ hold from 4 and inference rule (1).

The whole set of literals is then:

$$\{\sim b, \sim c, \neg a, a, \sim a, \sim \neg a, d, e, \sim d, \sim e, \sim \neg d, \sim \neg e\}.$$

N. B. We extend the language with the special symbol $\perp$. For every pair of classical literals $\{L, \neg L\}$ in the language of P we implicitly assume a rule $\perp \leftarrow L, \neg L$[7].

---

[6] We treat contradictory programs extending the approach of (Pereira et al. 1991a, Pereira et al. 1991b).

[7] This is not strictly necessary but simplifies the exposition. Furthermore, without loss of generality, we only consider rules $\perp \leftarrow L, \neg L$ for which rules for both L and $\neg L$ exist in P. We also use the notation $\perp_L$ to denote the head of rule $\perp \leftarrow L, \neg L$.

Definition 9 Pseudo–interpretation. A pseudo–interpretation (or p–interpretation for short) is a possibly contradictory set of ground literals from the language of a program.

We extend the Θ operator (Przymusinska and Przymusinski 1990) from the class of interpretations to the class of p–interpretations, and we call this the Θ^x operator (x standing for eXtended).

Definition 10 The Θ^x operator. Let P be a logic program and J a p–interpretation. The operator $\Theta_J^x : \mathcal{I} \to \mathcal{I}$ on the set $\mathcal{I}$ of all 3-valued p–interpretations of P is defined as follows: If $I \in \mathcal{I}$ is a p–interpretation of P and A is a ground classical literal then $\Theta_J^x(I)$ is the p–interpretation defined by:

1. $\Theta_J^x(I)(A) = 1$ iff there is a rule $A \leftarrow L_1, \ldots, L_n$ in P such that for all $i \leq n$ either $\hat{J}(L_i) = 1$, or L_i is positive and $I(L_i) = 1$;
2. $\Theta_J^x(I)(A) = 0$ iff one of the following holds:
   (a) for every rule $A \leftarrow L_1, \ldots, L_n$ in P there is an $i \leq n$, such that either $\hat{J}(L_i) = 0$, or L_i is positive and $I(L_i) = 0$;
   (b) $\hat{J}(\neg A) = 1$;
3. $\Theta_J^x(I)(A) = 1/2$ otherwise.

Note that the only difference between this definition and the definition of Θ operator introduced in (Przymusinska and Przymusinski 1990) is condition (2b) capturing the coherence requirement, or inference rule (1). Furthermore, since it is defined over the class of p–interpretations, it allows that for a given literal L, we may have $\Theta_J^x(I)(L) = 1$ as well as $\Theta_J^x(I)(L) = 0$.

Proposition 11. For every p–interpretation J, the operator Θ_J^x is monotone and has a unique least fixed point given by $\Theta_J^x{\uparrow^w}$, also denoted by $\Omega^x(J)$[8].

Definition 12 p–model. Given a program, a p–model is a p–interpretation I such that:

$$I = \Omega^x(I) \tag{2}$$

Remark. Note that if a p–model M is contradictory w.r.t. to L then M is inconsistent w.r.t. to L by virtue of inference rule (1), although the converse is not true.

Definition 13 Well Founded Model. The pseudo Well Founded Model M_P of P is the F–least p–model.

The non–minimal models satisfying (2) above are the (pseudo) Extended Models (XMs for short). To compute the p–model M_P, we define the following transfinite sequence $\{I_\alpha\}$:

---

[8] Recall (Przymusinska and Przymusinski 1990) that the F-least interpretation used to compute the least fixed point of $\Theta_J^x{\uparrow^w}$ is $\sim\mathcal{H}_P$.

$$I_0 = \langle \emptyset, \emptyset \rangle$$
$$I_{\alpha+1} = \Omega^x(I_\alpha) = \Theta_{I_\alpha}^{x\,\uparrow w}$$
$$I_\delta = \bigcup_{\alpha<\delta} I_\alpha \text{ for limit ordinal } \delta$$

Equivalently, the pseudo well founded model M_P of P is the F–least fixed point of (2) and is given by $M_P = I_\lambda = \Omega^{x\,\uparrow\lambda}$.

Definition 14. A program P is contradictory iff $\perp \in M_P$.

Example 7. Recall the program of example 6:

$$P = \{a \leftarrow \sim b \;;\; \neg a \leftarrow \sim c \;;\; d \leftarrow a \;;\; e \leftarrow \neg a\}.$$

$$\Theta_{I_0}^{\uparrow 0} = \{\sim a, \sim \neg a, \sim b, \sim \neg b, \sim c, \sim \neg c, \sim d, \sim \neg d, \sim e, \sim \neg e\}$$
$$\Theta_{I_0}^{\uparrow 3} = \Theta_{I_0}(\Theta_{I_0}^{\uparrow 2}) = \{\sim b, \sim c, \sim \neg b, \sim \neg c, \sim \neg d, \sim \neg e\} = \Theta_{I_0}^{\uparrow w} = I_1$$
$$\Theta_{I_1}^{\uparrow 3} = \Theta_{I_1}(\Theta_{I_1}^{\uparrow 2}) = \{a, d, \neg a, e, \sim b, \sim \neg b, \sim c, \sim \neg c, \sim \neg d, \sim \neg e\} = \Theta_{I_1}^{\uparrow w} = I_2$$
$$\Theta_{I_2}^{\uparrow 2} = \Theta_{I_2}(\Theta_{I_2}^{\uparrow 1}) = \{a, \sim \neg a, \neg a, \sim a, d, \sim \neg d, e, \sim \neg e, \sim b, \sim \neg b, \sim c, \sim \neg c, \sim d, \sim \neg d, \sim e\} =$$
$$= \Theta_{I_2}^{\uparrow w} = I_3$$
$$\Theta_{I_3}^{\uparrow 2} = \Theta_{I_3}(\Theta_{I_3}^{\uparrow 1}) = \{a, \sim \neg a, \neg a, \sim a, d, \sim \neg d, e, \sim \neg e, \sim b, \sim \neg b, \sim c, \sim \neg c, \sim d, \sim \neg d, \sim e\} =$$
$$= \Theta_{I_3}^{\uparrow w} = I_4 = I_3$$

so the program is contradictory.

4.2 Removing the contradiction

In order to get revised non–contradictory consistent models we must know where a contradiction arises. In this section we identify sets of default literals true by CWA whose revision to undefined can remove contradiction, by withdrawing the support of the CWAs on which the contradiction depends.

Definition 15 Dependency set. A Dependency Set of a literal L in a program P, represented as $DS(L)$, is obtained as follows:

1. If L is a classical literal:
   (a) if there are no rules for L then the only $DS(L) = \{L\}$.
   (b) for each rule $L \leftarrow B_1, \ldots, B_n(n \geq 0)$ in P for L, there exists one $DS_k(L) = \{L\} \cup \bigcup_i DS_{j(i)}(B_i)$ for each different combination k of one $j(i)$ for each i.
2. For a default literal $\sim L$:
   (a) if there are no rules in P for L then a $DS(\sim L) = \{\sim L\}$.
   (b) if there are rules for L then choose from every rule for L a single literal. For each such choice there exist several $DS(\sim L)$; each contains $\sim L$ and one dependency set of each default complement[9] of the chosen literals.

---

[9] The default complement of a classical literal L is $\sim L$; that of a default literal $\sim L$ is L.

(c) if there are rules for $\neg L$ then there are, **additionally**, dependency sets
$DS(\sim L) = \{\sim L\} \cup DS_k(\neg L)$ for each k.

Example 8. $P = \{a \longleftarrow \sim b \; ; \neg a \longleftarrow \sim c \; ; d \leftarrow a \; ; e \leftarrow \neg a\}$. In this case we have the following dependency sets:

$$
\begin{aligned}
DS(\sim b) &= \{\sim b\} & DS_1(\sim a) &= \{\sim a, b\} \\
DS(\sim c) &= \{\sim c\} & DS_2(\sim a) &= \{\sim a, \neg a, \sim c\} \\
DS(a) &= \{a, \sim b\} \\
DS(\neg a) &= \{\neg a, \sim c\} \\
DS_1(\sim d) &= \{\sim d\} \cup DS_1(\sim a) &= \{\sim d, \sim a, b\} \\
DS_2(\sim d) &= \{\sim d\} \cup DS_2(\sim a) &= \{\sim d, \sim a, \neg a, \sim c\} \\
DS(\perp_a) &= \{\perp_a, a, \neg a, \sim b, \sim c\}
\end{aligned}
$$

Definition 16 Support of a literal. A support $SS_M(L)$ w.r.t. to a model M is a non–empty dependency set $DS(L)$ such that $DS(L) \subseteq M$. If there exists a $SS_M(L)$ we say that L is supported in M.

For simplicity, a support w.r.t. the pseudo WFM M_P of P can be represented by $SS(L)$.

Definition 17 Support of a set of literals. A support w.r.t. to a model M is:

$$SS_{Mk}(\{L_1, \ldots, L_n\}) = \bigcup_i SS_{Mj(i)}(L_i)$$

For each combination k of $j(i)$ there exists one support of sets of literals.

With the notion of support we are able to identify which literals support a contradiction, i.e., the literal $\perp$. In order to remove a contradiction we must change the truth value of at least one literal from each support set of $\perp$. One issue is for which literals we allow to initiate a change of their truth values; another is how to specify a notion of minimal change.

As mentioned before, we only wish to initiate revision on default literals true by CWA. To identify such *revising* literals we first define:

Definition 18 Default support. A default literal $\sim A$ is default supported w.r.t. M if **all** supports $SS_M(\sim A)$ have only default literals.

Example 9. Let $P = \{\neg a; \; a \longleftarrow \sim b; \; b \leftarrow c; \; c \leftarrow d\}$. The only support of $\perp$ is $\{\neg a, a, \sim b, \sim c, \sim d\}$, and default supported literals are $\sim b$, $\sim c$, and $\sim d$. Here we are not interested in revising the contradiction by undefining $\sim b$ or $\sim c$ because they depend on $\sim d$. The reason is that we are attempting to remove only those contradictions based on $CWAs$. Now, the CWA of a literal that is supported on another depends on the CWA of the latter.

In order to make more precise what we mean we first present two definitions:

Definition 19 Self supported set. A set of default literals S is self supported w.r.t. a model M iff there exists a $SS_M(S) = S$.

Definition 20 Revising and co–revising literals. Given a program P with pseudo well-founded model M_P, we define co–$\mathcal{R}_P$, the *co–revising literals* induced by P, as the set of literals belonging to some minimal self supported set w.r.t. M_P. We define $\mathcal{R}_P$, the *revising literals*, as the set of co–revising literals L such that $\neg L \notin M_P$. The next examples motivate these definitions.

Example 10. Let $P = \{\neg p; \quad p \leftarrow \sim a; \quad \neg a \leftarrow \sim b\}$.

The co-revising literals are $\{\sim a, \sim b\}$ and the revising are $\{\sim b\}$. The difference is that to revise $\sim a$, one needs to change the truth value of $\neg a$ as well, in order to maintain coherence. To revise $\sim b$, there is no such need. Revising $\neg a$ only is not enough since then $\sim a$ becomes true by default.

Example 11. In the program of example 9, the self supported sets w.r.t. M_P are $\{\sim b, \sim c, \sim d\}$, $\{\sim c, \sim d\}$, and $\{\sim d\}$. Thus the only revising literal is $\sim d$. Note how the requirement of minimality ensures that only CWA literals not depending on other CWAs are revising. In particular:

Proposition 21.

1. If there are no rules for L, then $\sim L$ is a co–revising literal.
2. If there are no rules for L nor for $\neg L$, then $\sim L$ is a revising literal.
3. If $\sim L$ is co–revising and default supported, then it is revising.

An atom can also be false by CWA if it is involved in a positive *"loop"*. Such cases are also accounted for:

Example 12. Let P_1 and P_2 be:

$$P_1 = \{\neg a; \ a \leftarrow \sim b; \ b \leftarrow b, c\}$$
$$P_2 = \{\neg a; \ a \leftarrow \sim b; \ b \leftarrow b; \ b \leftarrow c\}$$

For P_1 self supported sets are: $\{\sim b, \sim c\}$, $\{\sim b\}$, and $\{\sim c\}$. Thus $\sim b$ and $\sim c$ are revising. For P_2 the only minimal self supported set is $\{\sim c\}$ thus only $\sim c$ is revising. The only support set of $\sim b$ is $\{\sim b, \sim c\}$. In P_2 it is clear that $\sim b$ depends on $\sim c$. So $\sim b$ is not revising. In P_1 the truth of $\sim b$ can support itself. Thus $\sim b$ is also revising.

Another class of literals is needed in the sequel. Informally, indissociable literals are those that strongly depend on each other, so that their truth value must always be the same. It is impossible to change the truth value of one without changing the truth value of another. So:

Definition 22 Indissociable set of literals. A set of default literals S is indissociable iff

$$\forall \sim a, \sim b \in S \quad \sim a \in \bigcap_i SS_i(\sim b) \ .$$

i.e., each literal in S belongs to every support of every literal in S.

Proposition 23. If S is a minimal self supported set and the only $SS(S)$ is S, then S is an indissociable set of literals.

Example 13. In P below, $\{\sim a, \sim b, \sim c\}$ is a set of indissociable literals:

$$\neg p \leftarrow \qquad\qquad a \leftarrow b$$
$$p \leftarrow \sim a \qquad\qquad b \leftarrow c$$
$$c \leftarrow a$$

Example 14. Let P be:

$$\neg p \leftarrow \qquad\qquad a \leftarrow b \qquad\qquad a \leftarrow c$$
$$p \leftarrow \sim a \qquad\qquad b \leftarrow a$$

We have:

$$SS(\sim c) = \{\sim c\}$$
$$SS(\sim b) = \{\sim a, \sim b, \sim c\}$$
$$SS(\sim a) = \{\sim a, \sim b, \sim c\}$$

and the unique indissociable set of literals is $\{\sim c\}$, which is also the set of revising literals.

Given the revising literals we have to find one on which the contradiction rests. This is done by finding the supports of $\perp$ where revising literals occur only as leaves (these constitute the $\perp$ assumption sets):

Definition 24 Assumption set. Let P be a program with (pseudo) WFM M_P and $L \in M_P$. An assumption set $AS(L)$ is defined as follows, where $\mathcal{R}_P$ is the set of the revising literals induced by P :

1. If L is a classical literal:
   (a) if there is a fact for L then the only $AS(L) = \{\}$.
   (b) for each rule $L \leftarrow B_1, \ldots, B_n (n \geq 1)$ in P for L such that $\{B_1, \ldots, B_n\} \subseteq M_P$, there exists one $AS_k(L) = \bigcup_i AS_{j(i)}(B_i)$ for each different combination k of one $j(i)$ for each i.
2. For a default literal $\sim L$:
   (a) if $\sim L \in \mathcal{R}_P$, then the only $AS(\sim L) = \{\sim L\}$.
   (b) if $\sim L \in co - \mathcal{R}_P$, then there is a $AS(\sim L) = \{\sim L\}$.
   (c) if $\sim L \notin co - \mathcal{R}_P$, then choose from every rule for L a single literal whose default complement belongs to M_P. For each such choice there exist several $AS(\sim L)$; each contains one assumption set of each default complement of the chosen literals.
   (d) if $\neg L \in M_P$, then there are, **additionally**, assumption sets $AS(\sim L) = AS_k(\neg L)$ for each k.

Definition 25. A program P is *revisable* iff no assumption set of $\perp$ is empty.

This definition entails that a program P is not revisable if $\perp$ has some support without co–revising literals.

Example 15. Consider $P = \{\neg a;\ a \longleftarrow \sim b\ ; b \longleftarrow \sim c\ ; c\}$ with

$$M_P = \{\bot, a, \sim a, \neg a, \sim b, c\}.$$

The only support of $\bot$ is $SS(\bot) = \{\bot, a, \sim a, \neg a, \sim b, c\}$. $co - \mathcal{R}_P = \{\}$. Thus $AS(\bot) = \{\}$ and the program is not revisable.

Definition 26 Removal set. A Removal Set (RS) of a literal L of program P is a set of literals formed by the union of one **non–empty** subset from each $AS_P(L)$.

Note that, although the program may induce revising literals, this is not enough for a program to be revisable.

In order to make minimal changes that preserve the indissociability of literals we define:

Definition 27 Minimal contradiction removal sets. Let R be a minimal removal set of $\bot$. A Minimal Contradiction Removal Set of program P is the smallest set $MCRS$ such that $R \subseteq MCRS$ and $MCRS$ is inclusive of indissociable literals.

Definition 28 Contradiction removal sets. A contradiction removal set (or CRS for short) of a program P is either a $MCRS$ or the union of $MCRS$s.

Example 16. Consider $P = \{a \longleftarrow \sim b\ ;\ b \longleftarrow \sim a\ ;\ \neg a\}$.

$$DS(\neg a) = \{\}\ \ DS(a) = \{\sim b\}\ \ DS(\sim b) = \{a\}$$

$$DS(\bot_a) = DS(a) \cup DS(\neg a) \supseteq \{\neg a, a, \sim b\}$$

The M_P is obtained as follows:

$$
\begin{aligned}
I_0 &= \emptyset & I_3 &= \{\neg a, \sim a, \sim \neg b, b\} = \Theta_{I_2}^x \uparrow^w \\
I_1 &= \{\neg a, \sim \neg b\} = \Theta_{I_0}^x \uparrow^w & I_4 &= \{\neg a, \sim a, \sim \neg b, b\} = \Theta_{I_3}^x \uparrow^w \\
I_2 &= \{\neg a, \sim a, \sim \neg b\} = \Theta_{I_1}^x \uparrow^w & I_5 &= I_4
\end{aligned}
$$

$M_P = \{\neg a, \sim a, b, \sim \neg b\}$ and $\bot_a \notin M_P$; thus the program is non–contradictory.

Example 17. Consider $P = \{a \longleftarrow \sim b\ ; \neg a \longleftarrow \sim c\ ; d \longleftarrow a\ ; e \longleftarrow \neg a\}$.

Since $\sim b$ and $\sim c$ are both revising literals $AS(\bot) = \{\sim b, \sim c\}$. The contradiction removal sets are:

$$
\begin{aligned}
CRS_1 &= RS_1(\bot_a) = \{\sim b\} \\
CRS_2 &= RS_2(\bot_a) = \{\sim c\} \\
CRS_3 &= RS_3(\bot_a) = \{\sim b, \sim c\}
\end{aligned}
$$

Example 18. Let P be:

$$a \leftarrow \sim b \qquad \neg a \leftarrow \sim f$$
$$\neg a \leftarrow \sim d \qquad \neg d \leftarrow \sim e$$

with $M_P = \{\sim b, \sim \neg b, \sim d, \sim e, \sim \neg e, \sim f, \sim \neg f, a, \neg a, \neg d, \sim a, \sim \neg a\}$.

Note that $\sim d$ is not co–revising, since there exists $SS(\sim d) = \{\sim d, \neg d, \sim e\}$. The revising literals are $\sim b$, $\sim e$, and $\sim f$. The assumption sets are:

$$AS_1(\bot) = \{\sim b, \sim e\}$$
$$AS_2(\bot) = \{\sim b\}$$
$$AS_3(\bot) = \{\sim b, \sim f\}$$

Thus the only contradiction removal set is $\{\sim b\}$.

Example 19. Consider the program $P = \{\neg a; a \leftarrow \sim d; \neg d \leftarrow \sim e\}$. We have $M_P = \{\sim e, \sim \neg e, \neg d, \sim d, a, \neg a, \sim \neg a, \sim a\}$ which is contradictory. The only revising literal is $\sim e$ and $\sim d$ is a co–revising literal. Hence one $AS(\sim d) = \{\sim d\}$ and the other $AS(\sim d) = \{\sim e\}$. The only CRS is $\{\sim d, \sim e\}$.

4.3 Contradiction free programs

Next we show that for each contradiction removal set there is a non-contradictory program obtained from the original one by a simple update. Based on these programs we define the $CRSX$ semantics.

Definition 29 CWA inhibition rule. The CWA inhibition rule for an atom A is $A \leftarrow \sim A$.

Any program P containing a CWA inhibition rule for atom A has no models containing $\sim A$.[10]

Definition 30 Contradiction free program. For each contradiction removal set CRS_i of a program P we engender the contradiction free program:

$$P_{CRS_i} =_{def} P \cup \{A \leftarrow \sim A| \sim A \in CRS_i\} \tag{3}$$

Proposition 31. For any contradiction removal sets i and j, $CRS_i \subseteq CRS_j \Rightarrow M_{P_{CRS_j}} \subseteq M_{P_{CRS_i}}$.

Theorem 32 Soundness of contradiction free programs. A contradiction free program P_{CRS} is non-contradictory, i.e., it has $WFSX$ semantics, and $M_{P_{CRS}} \subseteq M_P$.

---

[10] This rule can be seen as the *productive* integrity constraint $\leftarrow \sim A$. In fact, since the WF Semantics implicitly has in it the *productive* constraint $\leftarrow A, \sim A$, the inhibition rule can be seen as the minimal way of expressing by means of a program rule that $\sim A$ leads to an inconsistency, forcing A not to be false.

Example 20. Consider the program P :

$$a \leftarrow \sim b \qquad \neg c \leftarrow \sim d$$
$$b \leftarrow \sim a, \sim c \qquad c \leftarrow \sim e$$

The well founded model is

$$M_P = \{\perp_c, \sim d, \sim \neg d, \sim e, \sim \neg e, \neg c, c, \sim \neg c, \sim c, \sim b, \sim \neg b, a, \sim \neg a\}.$$

The contradiction removal sets are:

$$CRS_1 = \{\sim d\} \quad CRS_2 = \{\sim e\} \quad CRS_3 = \{\sim d, \sim e\}$$

with CRS_1 and CRS_2 being minimal w.r.t. set inclusion.

- $P_{CRS_1} = P \cup \{d \leftarrow \sim d\}$, with the unique model

$$M_P = \{\sim e, \sim \neg e, c, \sim \neg c, \sim b, \sim \neg b, a, \sim \neg a, \sim \neg d\}.$$

- $P_{CRS_2} = P \cup \{e \leftarrow \sim e\}$, with well founded model

$$M_P = \{\sim d, \neg c, \sim c, \sim \neg a, \sim \neg b, \sim \neg d\}.$$

- $P_{CRS_3} = P \cup \{e \leftarrow \sim e, d \leftarrow \sim d\}$ with well founded model

$$M_P = \{\sim \neg a, \sim \neg b, \sim \neg e, \sim \neg d\}.$$

Definition 33 $CRSX$ **Semantics.** Given a revisable contradictory program P, let CRS_i be any contradiction removal set for P. An interpretation I is a $CRSX$ model of P iff:

$$I = \Phi_{P_{CRS_i}}(I) . \tag{4}$$

The least (w.r.t. $\subseteq$) $CRSX$ model of P is called the $CRWFM$ model[11].

The contradiction removal semantics for logic programs extended with explicit negation is defined by the $WFSX$ well founded models of the revised programs defined by (4), representing the different forms of revising a contradictory program.

Example 21. For program P_1 of Example 12 the assumption sets are $AS_{1_1} = \{\sim b\}$ and $AS_{1_2} = \{\sim b, \sim c\}$. Thus the only CRS is $\{\sim b\}$, and the only $CRSX$ model is $\{\neg a, \sim a, \sim c\}$. For program P_2 the only assumption set is $AS_2 = \{\sim c\}$. Thus the only CRS is $\{\sim c\}$, and the only $CRSX$ model is $\{\neg a, \sim a\}$.

---

[11] This model always exists (cf. theorem 35.)

Example 22. Let P be:

$$a \leftarrow \sim b \qquad b \leftarrow c$$
$$\neg a \leftarrow \sim c \qquad c \leftarrow b$$

The only self supported set is $S = \{\sim b, \sim c\}$. Moreover the only support of S is itself. Thus $\sim b$ and $\sim c$ are revising and indissociable. As the only assumption set of $\perp$ is $\{\sim b, \sim c\}$ there are three removal sets: $\{\sim b\}$, $\{\sim c\}$, and $\{\sim b, \sim c\}$. Without indissociability one might think that for this program there would exist three distinct ways of removing the contradiction. This is not the case, since the $XSMs$ of P_{R_1}, P_{R_2}, and P_{R_3} are exactly the same, i.e., they all represent the same revision of P. This is accounted for by introducing indissociable literals in minimal contradiction removal sets. In fact there exists only one $MCRS$ $\{\sim b, \sim c\}$ and thus the only contradiction free program is P_{R_3}.

Theorem 34. For every i, j

$$CRS_i \neq CRS_j \Rightarrow WFM(P_{CRS_i}) \neq WFM(P_{CRS_j}) .$$

Theorem 35. The collection of contradiction free programs is a upper semi-lattice under set inclusion of rules in programs, and the set of revised models under set inclusion is a lower semi-lattice. There is a one–to–one correspondence between elements of both semi-lattices.

Acknowledgements

We thank Esprit BRA Compulog (no. 3012) and Compulog 2 (no. 6810), INIC and JNICT for their support.

References

Alferes, J.J., and L.M. Pereira: "On logic program semantics with two kinds of negation," in: K. Apt (ed.), *IJCSLP'92*, MIT Press (1992) 574-588.

Dung, P.M., and P. Ruamviboonsuk: "Well founded reasoning with classical negation," in: A. Nerode, W. Marek, and V. S. Subrahmanian (eds.), *LPNMR'91*, MIT Press (1991) 120-132.

Gelfond, M., and V. Lifschitz: "The stable model semantics for logic programming," in: R. A. Kowalski and K. A. Bowen (eds.), *5th ICLP*, MIT Press (1988) 1070-1080.

Gelfond, M., and V. Lifschitz: "Logic programs with classical negation," in: Warren and Szeredi (eds.), *ICLP*, MIT Press (September 1990) 579-597.

Inoue, K.: "Extended logic programs with default assumptions," in: Koichi Furukawa (ed.), *ICLP'91*, MIT Press (1991) 490-504.

Kowalski, R.: "Problems and promises of computational logic," in: John Lloyd (ed.), *Computational Logic Symposium*, Springer-Verlag (1990) 1-36.

Kowalski, R., and F. Sadri: "Logic programs with exceptions," in: Warren and Szeredi (eds.), *ICLP*, MIT Press (1990) 598-613.

Lloyd, J.W.: *Foundations of Logic Programming*. Symbolic Computation. Springer-Verlag, 1984.

Pereira, L.M. and J.J. Alferes: "Well founded semantics for logic programs with explicit negation," in: B. Neumann (ed.), *ECAI'92*, John Wiley & Sons, Ltd. (1992) 102-106.

Pereira, L.M., J.J. Alferes, and J.N. Aparício: "Contradiction Removal within Well Founded Semantics," in: A. Nerode, W. Marek, and V. S. Subrahmanian 102-106. (eds.) *LPNMR*, MIT Press (1991a) 105-119.

Pereira, L.M., J.J. Alferes, and J.N. Aparício: "The extended stable models of contradiction removal semantics," in: P.B arahona, L.M. Pereira, and A. Porto (eds.) *EPIA'91*, LNAI 541, Springer-Verlag (1991b) 105-119.

Pereira, L.M., J.J. Alferes, and J.N. Aparício: "Default theory for well founded semantics with explicit negation," in: D. Pearce and G. Wagner (eds.) *Logics for AI - JELIA'92*, LNAI 633, Springer-Verlag (1992a) 339-356.

Pereira, L.M., J. J. Alferes, and J. N. Aparício: "Well founded semantics with explicit negation and default theory," Technical report, AI Centre, Uninova, March 1992 (Submitted).

Pereira, L.M., J.N. Aparício, and J.J. Alferes: "Counterfactual reasoning based on revising assumptions," in: V. Saraswat and K. Ueda (eds.), *ILPS*, MIT Press, (1991c), 566-580.

Pereira, L.M., J.N. Aparício, and J.J. Alferes: "Hypothetical reasoning with well founded semantics," in: B.Mayoh (ed.), *Third Scandinavian Conf. on AI.* IOS Press, 1991d.

Pereira, L.M., J.N. Aparício, and J.J. Alferes: "Nonmonotonic reasoning with well founded semantics," in: K.Furukawa (ed.), *ICLP91*, MIT Press (1991e) 475-489.

Pereira, L.M., J.N. Aparício, and J.J. Alferes: "Non-monotonic reasoning with logic programming," in: *Journal of Logic Programming. Special issue on Nonmonotonic reasoning*, 1993 (to appear).

Pereira, L.M., J.N. Aparício, and J.J. Alferes: "Logic Programming for Non–Monotonic Reasoning." *In this volume*

Przymusinska, H., and T. Przymusinski: "Semantic issues in deductive databases and logic programs," in: R. Banerji (ed.), *Formal Techniques in Artificial Intelligence.* North Holland, 1990.

Przymusinski, T.: "Extended stable semantics for normal and disjunctive programs," in: Warren and Szeredi (eds.), *ICLP'90*, MIT Press (1990) 459-477.

Przymusinski, T.C.: "A semantics for disjunctive logic programs," in: D. Loveland, J. Lobo, and A. Rajasekar (eds.), *ILPS Workshop on Disjunctive Logic Programs*, 1991.

VanGelder, A., K.A. Ross, and J.S. Schlipf: "The well-founded semantics for general logic programs," in: *Journal of the ACM*, 38 (3) (1991) 620-650.

Wagner, G.: "A database needs two kinds of negation," in: B. Thalheim, J. Demetrovics, and H-D. Gerhardt (eds.), *MFDBS'91*, Springer-Verlag (1991) 357-371.

Logic Programming for Non–Monotonic Reasoning

Luís Moniz Pereira, Joaquim N. Aparício, José J. Alferes

CRIA, Uninova and DCS, U. Nova de Lisboa
2825 Monte de Caparica, Portugal
E-mail: {lmp/jna/jja}@fct.unl.pt

Abstract. Our purpose is to develop a modular systematic method of representing nonmonotonic reasoning problems with the Well Founded Semantics of extended logic programs augmented with eXplicit negation (WFSX), and augmented by its Contradiction Removal Semantics (CRSX) when needed. We show how to cast in the language of such logic programs forms of non-monotonic reasoning like defeasible reasoning and hypothetical reasoning, and apply them to different domains of knowledge representation, for instance taxonomic hierarchies and reasoning about actions. We then abstract a modular systematic method of representing non-monotonic problems in logic programming.

1 Introduction

Recently, several authors have showed the importance of introducing an explicit second kind of negation within logic programs, in particular for use in deductive databases, knowledge representation, and nonmonotonic reasoning (Dung and Ruamviboonsuk 1991, Gelfond and Lifschitz 1990, Inoue 1991, Kowalski and Sadri 1990, Kowalski 1990, (Pereira et al., 1991a, 1991b, 1991c), Wagner 1991, Przymusinski 1991).

It has been argued (Pereira et al., 1991a, 1991b, 1991c, 1992b) that well founded semantics are adequate to capture nonmonotonic reasoning if we interpret the least model provided by the semantics (called the Well Founded Model (*WFM*) of a program) as the skeptical view of the world, and the other models (called Extended Stable (*XM*) Models) as alternative enlarged consistent belief sets. A consequence of the well founded property is that the intersection of all models is itself a model belonging to the semantics.

Several proposals for extending logic programming semantics with a second kind of negation have been advanced. One such extension is the Answer Set (AS) semantics (Gelfond and Lifschitz 1990), which is an extension of Stable Model (SM) semantics (Gelfond and Lifschitz 1988) from the class of logic programs (Lloyd 1984) to those with a second form of negation. In (Kowalski and Sadri 1990) another proposal for such extension is introduced (based on the SM semantics), where implicitly a preference of negative information (exceptions) over positive information is assumed. However, AS semantics is not well founded. The meaning of the program is defined as the intersection of all answer-sets, and it is known that the computation of this intersection is computationally expensive. Another extension to include a second kind of negation is suggested by Przymusinski in (Przymusinski 1990). Although the set of models identified by this extension is well founded, the results are less intuitive (Alferes and Pereira 1992) with respect to the coexistence of both forms of negation. Based on the XSM semantics, Przymusinski (1991) also introduces a Stationary semantics where the second form of negation is classical negation. But classical negation entails that the logic programs under Stationary semantics no longer admit a procedural reading.

Well Founded Semantics with Explicit Negation (WFSX) (Pereira and Alferes 1992), is an extension of Well Founded Semantics WFS (VanGelder et al., 1990) including a second form of negation called *explicit negation*, which preserves the well founded property (cf. (Alferes and Pereira 1992) for a comparison of the above approaches) and admits procedural reading.

When a second form of negation is introduced, contradictions may be present (i.e., l and $\neg l$ hold for some l) and no semantics is given by WFSX[1]. In (Pereira et al., in this book) the authors define CRSX extending WFSX by introducing the notion of removing some contradictions and identifying the models obtained by revising closed world assumptions supporting those contradictions. One unique model, if any such revised model exists, is singled out as the contradiction free semantics. When no contradiction is present CRSX semantics reduces to WFSX semantics.

Here, using CRSX (which is assumed (Pereira et al., in this book)), we show how to cast in the language of logic programs extended with explicit negation different forms of non-monotonic reasoning such as defeasible reasoning and hypothetical reasoning, and apply it to diverse domains of knowledge representation such as taxonomic hierarchies and reasoning about actions.

Our final purpose is to identify a modular systematic method of representing some nonmonotonic reasoning problems with the CRSX semantics of logic programs.

2 Defeasible Reasoning

In this section we show how to represent defeasible reasoning with logic programs extended with explicit negation. We want to express defeasible reasoning and give a meaning to sets of rules (some of them being defeasible) when contradictions arise from the application of defeasible rules. For instance, we want to represent defeasible rules such as *birds normally fly* and *penguins normally don't fly*. Given a penguin, which is a bird, we adopt the skeptical point of view and none of the conflicting rules applies. Later on we show how to express preference for one rule over another in case they conflict and both are applicable. Consider for the moment a simpler version of this problem:

---

[1] In (Pereira et al., 1992a, 1992b) it is shown how WFSX relates to default theory.

Example 1. Consider the statements:

(*i*) *Normally birds fly.* (*ii*) *Penguins don't fly.* (*iii*) *Penguins are birds.* (*iv*) *a is a penguin.*

represented by the program P (with obvious abreviations, where *ab* stands for abnormal [2]):

$$f(X) \leftarrow b(X), \sim ab(X) \text{ (i)} \quad b(X) \leftarrow p(X) \text{ (iii)}$$
$$\neg f(X) \leftarrow p(X) \qquad\qquad \text{(ii)} \quad p(a) \qquad\qquad \text{(iv)}$$

Since there are no rules for $ab(a)$, $\sim ab(a)$ holds and $f(a)$ follows. On the other hand, we have $p(a)$, and $\neg f(a)$ follows from rule (ii). Thus the program model $\mathcal{M}_P$ is contradictory. In this case we argue that the first rule gives rise to a contradiction depending on a CWA on $ab(a)$ and so must not conclude $f(a)$. The intended meaning requires $\neg f(a)$ and $\sim f(a)$. We say that in this case a revision occcurs in the CWA of predicate instance $ab(a)$, which must turn to undefined. $\sim f(a)$ follows from $\neg f(a)$ in the semantics.

In this case CRSX identifies one contradiction removal set $CRS = \{\sim ab(a)\}$. The corresponding contradiction free program is $P \cup \{ab(a) \leftarrow \sim ab(a)\}$, and the corresponding model $CRWFM = \{p(a), \sim \neg p(a), b(a), \sim \neg b(a), \neg f(a), \sim f(a), \sim \neg ab(a)\}$.

In the example above the revision process is simple and the information to be revised is clearly the CWA about the abnormality predicate, so something can be said about *a* flying. However, this is not always the case, as shown in the following example:

Example 2. Consider the statements represented by *P* below:

(*i*) *Normally birds fly.* (*ii*) *Normally penguins don't fly.* (*iii*) *Penguins are birds.*

$$f(X) \leftarrow b(X), \sim ab_1(X) \text{ (i)}$$
$$\neg f(X) \leftarrow p(X), \sim ab_2(X) \text{ (ii)}$$
$$b(X) \leftarrow p(X) \qquad\qquad \text{(iii)}$$

Consider a penguin *a*, a bird *b*, and a rabbit *c* which does not fly; i.e., the facts are: $F = \{p(a), b(b), r(c), \neg f(c)\}$.

Remark. Facts *F* above and rule (*iii*) in *P* play the role of non–defeasible information, and should hold whichever world view one may choose for the interpretation of *P* together with those facts.

---

[2] $\sim$ is used for negation instead of the more usual *not* for expressing implicit or default negation (cf. failure to prove).

– W.r.t. the bird b everything is well defined and we have:

$$\left\{ \begin{array}{c} \sim ab_1(b), b(b), f(b), \sim\neg b(b), \sim\neg f(b), \sim\neg ab_1(b), \\ \sim\neg ab_2(b), \sim ab_2(b), \sim p(b), \sim\neg p(b) \end{array} \right\}$$

which says that bird b flies, $f(b)$, and it can't be shown it is a penguin, $\sim p(b)$. This is the intuitive result, since we may believe that b flies (because it is a bird) and it is not known to be a penguin, and so rules (i) and (ii) are non–contradictory w.r.t. bird b.

– W.r.t. the penguin a use of rules (i) and (ii) provoke a contradiction in $\mathcal{M}_P$: by rule (i) we have $f(a)$ and by rule (ii) we have $\neg f(a)$. Thus nothing can be said for sure about a flying or not, and the only non–ambiguous conclusions we may infer are:

$$\{ p(a), b(a), \sim\neg p(a), \sim\neg b(a), \sim\neg ab_1(a), \sim\neg ab_2(a) \}$$

Note that we are being skeptical w.r.t. $ab_1(a)$ and $ab_2(a)$, whose negation by CWA would give rise to a contradiction.

– W.r.t. c rules (i) and (ii) do not give rise to contradiction since $\sim p(c)$ and $\sim b(c)$ both hold, and we have:

$$\left\{ \begin{array}{c} r(c), \neg f(c), \sim p(c), \sim b(c), \sim\neg r(c), \sim f(c), \sim\neg p(c), \sim\neg b(c), \\ \sim\neg ab_1(c), \sim\neg ab_2(c), \sim ab_1(c), \sim ab_2(c) \end{array} \right\}$$

The least p–model of $P' = P \cup F$ using WFSX is[3]:

$$\begin{array}{l} \mathcal{M}_{P'} = \{ \, p(a), b(a), \sim\neg p(a), \sim\neg b(a), \sim\neg ab_1(a), \sim\neg ab_2(a), \sim ab_2(a), \\ \quad \sim\neg f(a), \sim f(a), \neg f(a), \sim ab_1(a), f(a), \\ \quad b(b), \sim\neg b(b), \sim\neg p(b), \sim\neg ab_1(b), \sim\neg ab_2(b), \sim ab_1(b), \sim ab_2(b), \\ \quad \neg f(c), \sim f(c), \sim p(c), \sim\neg p(c), \sim\neg b(c), \sim\neg f(c), \sim b(c), \\ \quad \sim\neg ab_1(c), \sim\neg ab_2(c), \sim ab_1(c), \sim ab_2(c) \} \end{array}$$

A contradiction arises concerning penguin a ($f(a)$ and $\neg f(a)$ both hold) because of the (closed world) assumptions on $ab_1(a)$ and $ab_2(a)$; the contradiction removal set is $CRS = \{\sim ab_1(a), \sim ab_2(a)\}$. Using CRSX we can determine formally the CRS above.

2.1 Exceptions

The notion of exception may be expressed in two different ways.

---

[3] Note that the difference between the $\mathcal{M}'_P$ model presented and the set of literals considered as the intuitive result in the previous remark differ precisely in the truth valuation of predicate instances $ab_1(a)$, $ab_2(a)$ and $f(a)$.

2.1.1. Exceptions to predicates.

Example 3. We express that the rule $flies(X) \leftarrow bird(X)$ applies whenever possible but can be defeated by exceptions using the rule:

$$flies(X) \leftarrow bird(X), \sim ab(X) \qquad (1)$$

If there is a bird b and a bird a which is known not to fly (and we don't know the reason why) we may express it by $\neg flies(a)$. In this case $\neg flies(a)$ establishes an *exception to the conclusion predicate* of the defeasible rule, and the meaning of the program[4] is:

$$\left\{ \begin{array}{l} bird(b), \sim ab(b), \sim \neg ab(b), \sim \neg bird(b), \sim \neg flies(b), flies(b) \\ bird(a), \qquad \sim \neg ab(a), \sim \neg bird(a), \neg flies(a), \sim flies(a) \end{array} \right\}$$

Note that nothing is said about $ab(a)$, i.e., the CWA on $ab(a)$ is avoided ($\{\sim ab(a)\}$ is the CRS) since it would give rise to a contradiction on $flies(a)$. This is the case where we know that bird a is an exception to the *normally birds fly* rule, by taking the fact into account that it does not fly: $\neg flies(a)$.

2.1.2. Exceptions to rules.
A different way to express that a given fact is an exception is to say that a given rule must not be applicable to that fact. To state that an element is an exception to a specific rule rather than to its conclusion predicate (more than one rule may have the same conclusion), we state that the element is abnormal w.r.t. the rule, i.e., the rule is not applicable to the element: if element a is an exception we express it as $ab(a)$.

In general we may want to express that a given X is abnormal under certain conditions. This is the case, for example, where we want to express penguins are abnormal w.r.t. the flying birds rule above, as follows:

$$ab(X) \leftarrow penguin(X) \qquad (2)$$

Rule (2) together with the non–defeasible rule $bird(X) \leftarrow penguin(X)$ add that *penguins are birds which are abnormal w.r.t. flying*. Similarly of dead birds; i.e., $ab(X) \leftarrow bird(X), dead(X)$ adding that *dead birds are abnormal w.r.t. flying*. Alternatively, given $\neg flies(X) \leftarrow dead(X)$, the non–abnormality of dead bird a w.r.t. flying, i.e., $\sim ab(a)$, may not be consistently assumed since it leads to a contradiction regarding $flies(a)$ and $\neg flies(a)$.

A stronger form of exception may be used to state that any element of a type (say penguin as an element of the type bird), X is considered an exception unless one knows it explicitly not to be one: $ab(X) \leftarrow \sim \neg penguin(X)$. One cannot apply defeasible rule (1) unless X is known not to be a penguin.

---

[4] This is a simplified version of Example 1.

2.2 Exceptions to exceptions

In general we may extend the notion of exceptioned rules to exception rules themselves, i.e., exception rules may be defeasible. This will allow us to express an exception to the exception rule for birds to fly, and hence the possibility that an exceptional penguin may fly, or that a dead bird may fly. In this case we want to say that the exception rule is itself a defeasible rule:

$$ab(X) \leftarrow bird(X), dead(X), \sim ab\_deadbird(X)$$

2.3 Preferences among rules

We may express now preference between two rules, stating that if one rule may be used, this constitutes an exception to the use of the other rule:

Example 4.

$$f(X) \leftarrow b(X), \sim ab_1(X) \text{ (i)}$$
$$\neg f(X) \leftarrow p(X), \sim ab_2(X) \text{ (ii)}$$
$$b(X) \leftarrow p(X) \qquad \text{(iii)}$$

In some cases we want to apply the most specific information; above, there should be (since a penguin is a specific kind of bird) an explicit preference for the non–flying penguins rule over the flying birds rule:

$$ab_1(X) \leftarrow p(X), \sim ab_2(X) \qquad\qquad (3)$$

If we have also *penguin(a)* and *bird(b)* the unique model is:

$$\left\{ \begin{array}{c} \sim ab_2(b), b(b), \sim p(b), \sim \neg f(b), \sim ab_1(b), f(b), \\ \sim ab_2(a), p(a), b(a), ab_1(a), \sim f(a) \end{array} \right\}$$

Rule (3) says that if a given penguin is not abnormal w.r.t. non–flying then it must be considered abnormal w.r.t. flying. So we infer that b is a flying bird, a (being a penguin) is also a bird, and that there is no evidence that it flies $\sim f(a)$.

3 Representation of Hierarchical Taxonomies

In this section we illustrate how to represent taxonomies with logic programs with explicit negation. In this representation we wish to express general abso-lute (i.e., non–defeasible) rules, defeasible rules, exceptions to defeasible rules, as well as exceptions to exceptions, by explicitly making preferences among de-feasible rules. As we've seen, when defeasible rules contradict each other and no preference rule is present, none of them is considered applicable in the most skeptical reading. We want to be able to express preference for one defeasible rule over another whenever they conflict. In taxonomic hierarchies we wish to express that in the presence of contradictory defeasible rules we prefer the one with most specific information (e.g., for a penguin, which is a bird, we want to conclude that it doesn't fly).

Example 5. The statements, facts and preferences about the domain are:

(1) Mammals are animals. (6) Normally animals don't fly.
(2) Bats are mammals. (7) Normally bats fly.
(3) Birds are animals. (8) Normally birds fly.
(4) Penguins are birds. (9) Normally penguins don't fly.
(5) Dead animals are animals. (10) Normally dead animals don't fly.

(11) Pluto is a mammal. (12) Tweety is a bird.
(13) Joe is a penguin. (14) Dracula is a bat.
(15) Dracula is a dead animal.

(16) Dead bats do not fly though bats do.
(17) Dead birds do not fly though birds do.
(18) Dracula is an exception to the above preferences.

Our representation of the hierarchy is the program:

$$animal(X) \leftarrow mammal(X) \quad (1) \qquad mammal(pluto) \ (11)$$
$$mammal(X) \leftarrow bat(X) \quad (2) \qquad bird(tweety) \ (12)$$
$$animal(X) \leftarrow bird(X) \quad (3) \qquad penguin(joe) \ (13)$$
$$bird(X) \leftarrow penguin(X) \quad (4) \qquad bat(dracula) \ (14)$$
$$animal(X) \leftarrow dead\_animal(X) \quad (5) \ dead\_animal(dracula) \ (15)$$
$$\neg flies(X) \leftarrow animal(X), \sim ab_1(X) \quad (6)$$
$$flies(X) \leftarrow bat(X), \sim ab_2(X) \quad (7)$$
$$flies(X) \leftarrow bird(X), \sim ab_3(X) \quad (8)$$
$$\neg flies(X) \leftarrow penguin(X), \sim ab_4(X) \quad (9)$$
$$\neg flies(X) \leftarrow dead\_animal(X), \sim ab_5(X) \ (10)$$

with the implicit hierarchical preference rules (greater specificity):

$$ab_1(X) \leftarrow bat(X), \sim ab_2(X)$$
$$ab_1(X) \leftarrow bird(X), \sim ab_3(X)$$
$$ab_3(X) \leftarrow penguin(X), \sim ab_4(X)$$

and the explicit problem statement preferences:

$$ab_2(X) \leftarrow dead\_animal(X), bat(X), \sim ab_5(X) \quad (16)$$
$$ab_3(X) \leftarrow dead\_animal(X), bird(X), \sim ab_5(X) \quad (17)$$
$$ab_5(dracula) \quad (18)$$

As expected, this program has exactly one model (coinciding with the minimal WFSX model) which is non–contradictory, no choices are possible and everything is defined in the hierarchy.

Thus pluto doesn't fly, and isn't an exception to any of the rules; tweety flies because it's a bird and an exception to the "animals don't fly" rule; joe doesn't fly because it's a penguin and an exception to the "birds fly" rule.

Note that although dracula is a dead animal, which by default don't fly (cf. rule (10)) it is also considered an exception to this very same rule. Furthermore, rule (16) saying that "dead bats normally do not fly" is also exceptioned by dracula and thus the "bats fly" rule applies and dracula flies. Note that preferences rules must be present in order to prevent contradictions from arising.

4 Hypothetical Reasoning

In hierarchies complete taxonomies are given, leaving no choices available. This is not always the case in hypothetical reasoning.

4.1 Hypothetical facts and rules

In some cases we want to be able to hypothesize about the applicability of rules or facts. This is distinct from just not having any knowledge at all about rules or facts.

Hypothetical facts. Consider this simple example: *John and Nixon are quakers* and *John is a pacifist* represented by the program

$$P_1 = \{quaker(john) \ pacifist(john) \ quaker(nixon)\} \ .$$

The $\mathcal{M}_{P_1}$ (which is the only XM model) is:

$$\left\{ \begin{array}{l} quaker(nixon), \sim pacifist(nixon), \sim\neg quaker(nixon), \sim\neg pacifist(nixon), \\ quaker(john), \quad pacifist(john), \quad \sim\neg quaker(john), \quad \sim\neg pacifist(john) \end{array} \right\}$$

and expresses exactly what is intended, i.e., John and Nixon are quakers, John is a pacifist and we have no reason to believe Nixon is a pacifist, in this or any other model (there aren't any others in fact). Now suppose we want to add:

$$Nixon \ might \ be \ a \ pacifist. \tag{4}$$

In our view we wouldn't want in this case to be so strong as to affirm *pacifist(nixon)*, thereby not allowing for the possibility of Nixon not being a pacifist. What we are prepared to say is that Nixon might be a pacifist if we don't have reason to believe he isn't and, vice-versa, that Nixon might be a non–pacifist if we don't have reason to believe he isn't one. Statement (4) is thus expressed as:

$$pacifist(nixon) \leftarrow \sim\neg pacifist(nixon) \tag{5}$$

$$\neg pacifist(nixon) \leftarrow \sim pacifist(nixon) \tag{6}$$

The first rule states that Nixon is a pacifist if there is no evidence against it. The second rule makes a symmetric statement. Let P_2 be the program P together with these rules. P_2 has a minimal model $\mathcal{M}_{P_2}$ (which is non–contradictory):

$$\left\{ \begin{array}{ll} quaker(nixon), & \sim\neg quaker(nixon), \\ quaker(john), pacifist(john), & \sim\neg quaker(john), \sim\neg pacifist(john) \end{array} \right\}$$

and two more XMs: $XSM_1 = \mathcal{M}_{P_2} \cup \{pacifist(nixon), \sim\neg pacifist(nixon)\}$ and $XSM_2 = \mathcal{M}_{P_2} \cup \{\neg pacifist(nixon), \sim pacifist(nixon)\}$. which is the result we were seeking. Statements of the form of (4) we call *unknown possible facts*; they are expressed in the form of (5) and (6). They can be read as a fact and its negation, each of which can be assumed only if this is consistent to do so.

Hypothetical rules. Consider the well known Nixon–diamond example using hypothetical rules instead of defeasible ones.

We represent these rules as named rules (in the fashion of (Poole 1988)) where the rule name may be present in one model as true, and in others as false.

Normally quakers are pacifists.	Normally republicans are hawks.
Pacifists are non hawks.	Hawks are non pacifists.
Nixon is a quaker and a republican.	Pacifists are non hawks.
There are other republicans.	There are other quakers.

The corresponding logic program is:

$$pacifist(X) \leftarrow quaker(X), hypqp(X) \qquad quaker(nixon)$$
$$hypqp(X) \leftarrow \sim\neg hypqp(X) \qquad republican(nixon)$$
$$hawk(X) \leftarrow republican(X), hyprh(X) \qquad quaker(another\_quaker)$$
$$hyprh(X) \leftarrow \sim\neg hyprh(X) \qquad republican(another\_republican)$$
$$\neg hawk(X) \leftarrow pacifist(X)$$
$$\neg pacifist(X) \leftarrow hawk(X)$$

where the following rules are also added, making each normality instance rule about Nixon hypothetical rather than defeasible (c.f. the representation of defeasible rules in Sect. 2):

$$hypqp(nixon) \leftarrow\sim\neg hypqp(nixon) \qquad hyprh(nixon) \leftarrow\sim\neg hyprh(nixon)$$
$$\neg hypqp(nixon) \leftarrow\sim hypqp(nixon) \qquad \neg hyprh(nixon) \leftarrow\sim hyprh(nixon)$$

as represented in Fig. 1.

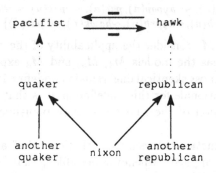

Fig. 1: The Nixon–diamond

The whole set of models is represented in Fig. 2.

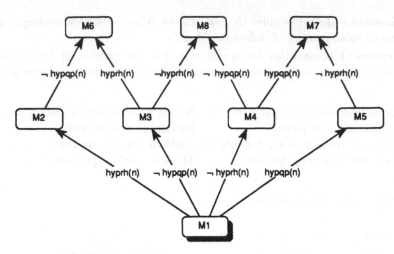

Fig. 2: Models of the Nixon–diamond problem using hypothetical rules, where edge labels represent the hypothesis being made when going from one model to another.

where the models (with obvious abbreviations) are:

$M_1 = \{qua(n), rep(n), \sim\neg qua(n), \sim\neg rep(n),$
$\quad qua(a\_qua), \sim\neg qua(a\_qua), \sim rep(a\_qua), \sim\neg rep(a\_qua),$
$\quad hypqp(a\_qua), \sim\neg hypqp(a\_qua), pac(a\_qua), \sim\neg pac(a\_qua),$
$\quad hyprh(a\_qua), \sim\neg hyprh(a\_qua), \sim\neg pac(a\_qua), \sim hawk(a\_qua)$
$\quad rep(a\_rep), \sim\neg rep(a\_rep), \sim qua(a\_rep), \sim\neg qua(a\_rep)$
$\quad hyprp(a\_rep), \sim\neg hyprp(a\_rep), rep(a\_rep), \sim\neg rep(a\_rep),$
$\quad hypqp(a\_rep), \sim\neg hypqp(a\_rep), \sim pac(a\_rep), \sim\neg hawk(a\_rep)\}$
$M_2 = M_1 \cup \{hyprh(n), \sim\neg hyprh(n), hawk(n), \sim\neg hawk(n), \neg pac(n), \sim pac(n)\}$
$M_3 = M_1 \cup \{\neg hypqp(n), \sim hypqp(n), \sim pac(n), \sim\neg hawk(n)\}$
$M_4 = M_1 \cup \{\neg hyprh(n), \sim hyprh(n), \sim hawk(n), \sim\neg pac(n)\}$
$M_5 = M_1 \cup \{hypqp(n), \sim\neg hypqp(n), pac(n), \sim\neg pac(n), \neg hawk(n), \sim hawk(n)\}$
$M_6 = M_2 \cup \{\neg hypqp(n), \sim hypqp(n), \sim pac(n), \sim\neg hawk(n)\}$
$M_7 = M_4 \cup \{hypqp(n), \sim\neg hypqp(n), pac(n), \sim\neg pac(n), \sim\neg hawk(n)\}$
$M_8 = M_3 \cup \{\neg hyprh(n), \sim hyprh(n), \sim hawk(n), \sim\neg pac(n)\}$

The models M_2 and M_6 consider the applicability of the *republicans are hawks* normality rule, whereas the models M_4, M_5, and M_8 explore not applying it. Model M_1, being the most skeptical one, remains undefined about the applicability of the rule. The rationale for this undefinedness is that since the application and the non–application of the rule are equally plausible, one should remain undecided about it.

Note here the distinction between "hypothetical rules" and "defeasible rules". While the latter are applied "whenever possible", that is, unless their applications lead to contradiction, the former provide equally plausible alternative extensions.

Remark. Note that with this form of representation we might as well add *abqp* or ¬*abqp*, explicit negative information and positive information are treated in the same way. In this case we may now hypothesize about the applicability and non–applicability of each normality rule. However, the most skeptical model (where no hypotheses are made) is still identical to the one where normality rules are interpreted as defeasible rules, the difference being that in the first case no revision is enforced since the $\mathcal{M}_P$ model is non–contradictory.

5 Application to Reasoning About Actions

We now apply the programing methodology used above to some reasoning about action problems and show that it gives correct results. The situation calculus notation (McCarthy and Hayes 1987) is used, where predicate $holds(P, S)$ expresses that property or fluent P holds in situation S; predicate $normal(P, E, S)$ expresses that in situation S, event or action E does not normally affect the truth value of fluent P; the term $result(E, S)$ names the situation resulting from the occurrence of event E in situation S.

5.1 The Yale Shooting Problem

This problem, supplied in (Hanks and Dermott 1986), will be represented in a form nearer to the one in (Kowalski and Sadri 1990).

Example 6. The problem and its formulation are as follows:
- Initially (in situation s0) a person is alive: $holds(alive, s0)$.
- After loading a gun the gun is loaded: $holds(loaded, result(load, S))$.
- If the gun is loaded then after shooting it the person will not be alive:

$$\neg holds(alive, result(shoot, S)) \leftarrow holds(loaded, S).$$

- After an event things normally remain as they were (frame axioms), i.e:
  - properties holding before the event will normally still hold afterwards:
    $$holds(P, result(E, S)) \leftarrow holds(P, S), \sim ab(P, E, S) \ (pp)^5$$
  - and properties not holding before the event will normally not hold afterwards:
    $$\neg holds(P, result(E, S)) \leftarrow \neg holds(P, S), \sim ab(\neg P, E, S) \ (np)^6$$

Consider the question "What holds and what doesn't hold after the loading of a gun, a period of waiting, and a shooting ?" represented as two queries:

$$\leftarrow holds(P, result(shoot, result(wait, result(load, s0))))$$
$$\leftarrow \neg holds(P, result(shoot, result(wait, result(load, s0))))$$

With this formulation the $\mathcal{M}_P$ model is the only XM Model. The subset of its elements that match with at least one of the queries is[7]:

$$\{holds(loaded, s3), \sim\neg holds(loaded, s3), \neg holds(alive, s3), \sim holds(alive, s3)\}$$

---

[5] pp stands for positive persistence.

[6] np stands for negative persistence; the use of $\neg$ as a functor permits a more compact representation of (pp) and (np), with a single ab predicate.

[7] Where s3 denotes the term result(shoot,result(wait,result(load,s0))).

which means that in situation s3 the gun is loaded and the person is not alive. This result coincides with the one obtained in (Kowalski 1990) for *holds*.

5.2 The Stolen Car Problem

The "Stolen Car Problem" (Kautz 1986) is expressed as follows:

1. I leave my car parked and go away: $holds(pk, s0)$.
2. I return after a while and the car is not there, i.e., after n wait events the car is not parked were I left it: $\neg holds(pk, s_n)$ where s_n represents the expression $r(w, \ldots, r(w, s0) \ldots)$ comprising n wait events.
3. After an event things normally remain as they were, represented as in the previous example.
4. In which situations S, between $s0$ and s_{n-1}, is the car parked and in which is it not?

Using CRSX the result agrees with common sense, and avoids the counterintuitive outcome that the car is parked in the situation just before I return, given by (Kautz 1986, Lifschitz 1986, Shoham 1988).

The WFSX assigns no meaning to the program since the pseudo *WFM* contains both $\neg holds(pk, s_n)$ (by 2) and $holds(pk, s_n)$ (by 1 and the frame axioms). The only assumption set of the contradiction is:

$$\{\sim ab(pk, w, s0), \sim ab(pk, w, s_1), \ldots, \sim ab(pk, w, s_{n-1}\}$$

showing that a contradiction results on assuming persistence at every wait event. So, there are n contradiction removal sets, namely:

$$CRS_1 = \{\sim ab(pk, w, s0)\}, \ldots, CRS_n = \{\sim ab(pk, w, s_{n-1})\} \ .$$

Accordingly:

- The $CRWFM$ is $\{holds(pk, s0), \neg holds(pk, s_n)\}$.
- There are n more $CRXSM$s:

$$CRXSM_1 = \{holds(pk, s0), holds(pk, s_1), \neg holds(pk, s_n)\}$$

$$\ldots$$

$$CRXSM_n = \{holds(pk, s0), \ldots, holds(pk, s_{n-1}), \neg holds(pk, s_n)\}$$

This can be interpreted in the following way:

- In the most skeptical view ($CRWFM$) one can only say for sure that the car was parked when I left it, and not parked when I returned. It is undefined whether the car was parked in all other intermediate situations.
- For each intermediate situation there is the possibility (a $CRXSM$) that the car was still parked.

Note how the set of $CRXSM$s can be read as the disjunct: either the car disappeared in situation s_1 or in situation s_2 or ... or in situation s_{n-1}.

6 Summary of Our Representation Method

In this section we summarize and systematize the representation method adopted in all the above examples. The type of rules for which we propose a representation is, in our view, general enough to capture a wide domain of nonmonotonic problems. Each type of rule is described in a subsection by means of a schema in natural language and its corresponding representation rule.

Definite Rules. *If A then B.* The representation is: $B \leftarrow A$.
Definite Facts. *A is true.* The representation is: A. *A is false.* The representation is: $\neg A$.
Defeasible (or maximally applicable) Rules. *Normally if A then B.* The representation is:

$$B \leftarrow A, \sim ab.$$

where $\sim ab$ is a new predicate symbol. As an example consider the rule "Normally birds fly". Its representation is: $fly(X) \leftarrow bird(X), \sim ab(X)$.
Defeasible Facts are a special case of *Defeasible Rules* where A is absent.
Known Exceptions to Defeasible Rules. *Under certain conditions COND there are known exceptions to the defeasible rule $H_1 \leftarrow B_1, \sim ab_1$.*

$$ab_1 \leftarrow COND.$$

As an example, the representation of the exception "Penguins are exceptions to the "normally birds fly" rule (i.e., rule $f \leftarrow b, \sim abb$)" is: $abb \leftarrow penguin$.
Possible Exceptions to Defeasible Rules. *Under certain conditions COND there are no possible exceptions to the defeasible rule $H_1 \leftarrow B_1, \sim ab_1$.*

$$ab_1 \leftarrow \sim \neg COND.$$

As an example, the representation of the exception "Animals not known to be not a penguin are exceptions to the "normally birds fly" rule (i.e., the rule $f \leftarrow b, \sim abb$)" is: $abb \leftarrow \sim \neg penguin$.
Preference rules are a special kind of exception to defeasible rules:
Preference Rules. *Under conditions COND, prefer to apply the defeasible rule $H_1 \leftarrow B_1, \sim ab_1$ instead of the defeasible rule $H_2 \leftarrow B_2, \sim ab_2$.*

$$ab_1 \leftarrow COND, \sim ab_2.$$

As an example consider "For penguins, if the rule that says "normally penguins don't fly" is applicable then inhibit the "normally birds fly" rule". This is represented as: $ab\_b \leftarrow penguin(X), \sim ab\_penguin(X)$.
Unknown Possible Fact. *F might be true or not* (in other words, the possibility or otherwise of F should be considered).

$$F \leftarrow \sim \neg F.$$
$$\neg F \leftarrow \sim F.$$

Hypothetical (or possibly applicable) Rules. *Rule "If A then B" may or may not apply.* Its representation is:

$$B \leftarrow A, hyp$$
$$hyp \leftarrow \sim\neg hyp$$
$$\neg hyp \leftarrow \sim hyp$$

where *hyp* is a new predicate symbol. As an example consider the rule. "Quakers might be pacifists". Its representation is:

$$pacifist(X) \leftarrow quaker(X), hypqp(X).$$

$$hypqp(X) \leftarrow \sim\neg hypqp(X).$$
$$\neg hypqp(X) \leftarrow \sim hypqp(X).$$

Acknowledgements

We thank ESPRIT BRA projects COMPULOG, COMPULOG II, INIC, JNICT, and Gabinete de Filosofia do Conhecimento for their support.

References

Alferes, J.J., and L.M. Pereira: "On logic program semantics with two kinds of negation," in: *IJCSLP'92*. MIT Press, 1992.

Dung, P.M., and P. Ruamviboonsuk: "Well founded reasoning with classical negation," in: A. Nerode, W. Marek, and V.S. Subrahmanian (eds.), *Workshop on LPNMR*. MIT Press, 1991.

Gelfond, M., and V. Lifschitz: "The stable model semantics for logic programming," in: R.A. Kowalski and K.A. Bowen (eds.), *5th ICLP*, MIT Press (1988) 1070-1080..

Gelfond, M., and V. Lifschitz: "Logic programs with classical negation," in: Warren and Szeredi (eds.), *ICLP*, MIT Press (1990) 579-597.

Hanks, S., and D. McDermott: "Default reasoning, nonmonotonic logics and the frame problem," in: *AAAI86* (1986) 328-333.

Inoue, K.: "Extended logic programs with default assumptions," in: K.Furukawa (ed.), *8th ICLP*, MIT Press (1991) 490-504.

Kautz, H.: "The logic of persistence," in: *AAAI'86* (1986) 401.

Kowalski, R.: "Problems and promises of computational logic," in: J.Lloyd (ed.), *Computational Logic Symposium* Springer-Verlag (1990) 1-36.

Kowalski, R., and F.Sadri: "Logic programs with exceptions," in: Warren and Szeredi (eds.), *ICLP*, MIT Press (1990) 598-613.

Lifschitz, V.: "Pointwise circumscription," in: *AAAI'86* (1986) 406.

Lloyd, J.W.: *Foundations of Logic Programming*. Symbolic Computation. Springer-Verlag, 1984.

McCarthy, J., and P.J. Hayes: *Some Philosophical Problems from the Standpoint of Artificial Intelligence*. Readings in Nonmonotonic Reasoning. M. Kaufmann Inc. (1987) 26-45.

Pereira, L.M., and J.J. Alferes: "Well founded semantics for logic programs with explicit negation," in: *ECAI'92*. John Wiley & Sons, Ltd, 1992.

Pereira, L.M., J.J. Alferes, and J.N. Aparício: "Contradiction removal semantics with explicit negation." *In this book.*

Pereira, L.M., J.J. Alferes, and J.N. Aparício: "Default theory for well founded semantics with explicit negation," in: *JELIA '92*, 1992a.

Pereira, L.M., J.J. Alferes, and J.N. Aparício: "Well founded semantics with explicit negation and default theory," Technical report, AI Center, Uninova, 1992b. *Submitted.*

Pereira, L.M., J.N. Aparício, and J.J. Alferes: "Counterfactual reasoning based on revising assumptions," in: V.Saraswat and K.Ueda (eds.), *ILPS*, MIT Press (1991a) 566-580.

Pereira, L.M., J.N. Aparício, and J.J. Alferes: "Hypothetical reasoning with well founded semantics," in: B.Mayoh (ed.) *Third Scandinavian Con. on AI.* IOS Press, 1991b.

Pereira, L.M., J.N. Aparício, and J.J. Alferes. "Nonmonotonic reasoning with well founded semantics," in: K. Furukawa (ed.) *ICLP91*, MIT Press (1991c) 475-489.

Poole, D.L.: "A logical framework for default reasoning," in: *Journal of AI*, 36 (1), (1988) 27-47.

Przymusinski, T.: "Extended stable semantics for normal and disjunctive programs," in: Warren and Szeredi (eds.) *ICLP90*, MIT Press (1990) 459-477.

Przymusinski, T.C.: "A semantics for disjunctive logic programs," in: D. Loveland, J. Lobo, and A. Rajasekar (eds.), *ILPS Workshop on Disjunctive Logic Programs*, 1991.

Shoham, Y.: *Reasoning about Change: Time and Change from the Standpoint of Artificial Intelligence.* MIT Press, 1988.

VanGelder, A., K.A. Ross, and J.S. Schlipf: "The well-founded semantics for general logic programs," in: *Journal of the ACM* (1990) 221-230.

Wagner, G.: "A database needs two kinds of negation," in: B. Thalheim, J. Demetrovics, and H-D. Gerhardt (eds.), *MFDBS'91*, Springer-Verlag (1991) 357-371.

Agent Oriented Programming:
An overview of the framework and summary of recent research

Yoav Shoham

Computer Science Department, Stanford University, Stanford, CA 94305
E-mail: shoham@flamingo.stanford.edu

Abstract. This is a short overview of the *agent oriented programming* (AOP) framework. AOP can be viewed as an specialization of *object-oriented programming*. The state of an agent consists of components called beliefs, choices, capabilities, commitments, and possibly others; for this reason the state of an agent is called its *mental state*. The mental state of agents is captured formally in an extension of standard epistemic logics: beside temporalizing the knowledge and belief operators, AOP introduces operators for commitment, choice and capability. Agents are controlled by *agent programs*, which include primitives for communicating with other agents. In the spirit of *speech-act theory*, each communication primitives is of a certain type: informing, requesting, offering, and so on. This paper describes these features in a little more detail, and summarizes recent results and ongoing AOP-related work.

1 Introduction

Agent Oriented Programming is a proposed new programming paradigm, based on a societal view of computation. Although new, the proposal benefits from extensive previous research. Indeed, the discussion here touches on issues that are the subject of much current research in AI, issues which include the notion of agenthood and the relation between a machine and its environment. Many of the ideas here intersect and interact with the ideas of others. In this overview, however, I will not place this work in the context of other work. That, as well as more details on AOP, appear in (Shoham 1993), and subsequent publications which are mentioned below.

1.1 What is an agent?

The term 'agent' is used frequently these days. This is true in AI, but also outside it, for example in connection with data bases and manufacturing automation. Although increasingly popular, the term has been used in such diverse ways that it has become meaningless without reference to a particular notion of agenthood. Some notions are primarily intuitive, others quite formal. In several longer publications I outline several senses of agenthood that I have discerned in the AI literature. Given the limited space, here I will directly present "my" sense of agenthood.

I will use the term '(artificial) agents' to denote entities possessing formal versions of mental state, and in particular formal versions of beliefs, capabilities, choices, commitments, and possibly a few other mentalistic-sounding qualities. What will make any hardware or software component an agent is precisely the fact that one has chosen to analyze and control it in these mental terms.

The question of what an agent is now replaced by the question of what can be described in terms of knowledge, belief, commitment, *et cetera*. The answer is that *anything* can be so described, although it is not always advantageous to do so. D. Dennett proposes the "intentional stance," from which systems are ascribed mental qualities such as intentions and free will (Dennett 1987). The issue, according to Dennett, is not whether a system really is intentional, but whether we can coherently view it as such. Similar sentiments are expressed by J. McCarthy in his 'Ascribing Mental Qualities to Machines' paper (McCarthy 1979), who also distinguishes between the 'legitimacy' of ascribing mental qualities to machines and its 'usefulness.' In other publications I illustrate the point through the light-switch example. It is perfectly coherent to treat a light switch as a (very cooperative) agent with the capability of transmitting current at will, who invariably transmits current when it believes that we want it transmitted and not otherwise; flicking the switch is simply our way of communicating our desires. However, while this is a coherent view, it does not buy us anything, since we essentially understand the mechanism sufficiently to have a simpler, mechanistic description of its behavior. In contrast, we do not have equally good knowledge of the operation of complex systems such as robots, people, and, arguably, operating systems. In these cases it is often most convenient to employ mental terminology; the application of the concept of 'knowledge' to distributed computation, discussed below, is an example of this convenience.

1.2 Agent- versus object-oriented programming

Adopting the sense of agenthood just described, I have proposed a computational framework called *agent oriented programming* (AOP). The name is not accidental, since from the engineering point of view AOP can be viewed as a specialization of the *object-oriented programming* (OOP) paradigm. I mean the latter in the spirit of Hewitt's original Actors formalism, rather than in some of the senses in which it used today. Intuitively, whereas OOP proposes viewing a computational system as made up of modules that are able to communicate with one another and that have individual ways of handling incoming messages, AOP specializes the framework by fixing the state (now called *mental state*) of the modules (now called *agents*) to consist of precisely-defined components called beliefs (including beliefs about the world, about themselves, and about one another), capabilities, choices, and possibly other similar notions. A computation consists of these agents informing, requesting, offering, accepting, rejecting, competing, and assisting one another. This idea is borrowed directly from the *speech act* literature. Speech-act theory categorizes speech, distinguishing between informing, requesting, offering and so on; each such type of communicative act involves different presuppositions and has different effects. Speech-act theory has been applied in AI, in natural language research as well as in plan recognition. To my knowledge, AOP and McCarthy's Elephant2000 language are the first attempts to base a programming language in part on speech acts. Fig. 1 summarizes the relation between AOP and OOP.[1]

Framework:	OOP	AOP
Basic unit:	object	agent
Parameters defining state of basic unit:	unconstrained	beliefs, commitments, capabilities, choices, ...
Process of computation:	message passing and response methods	message passing and response methods
Types of message:	unconstrained	inform, request, offer, promise, decline, ...
Constraints on methods:	none	honesty, consistency, ...

Fig. 1. OOP versus AOP

1.3   On the use of pseudo-mental terminology

The previous discussion referred to mentalistic notions such as belief and commitment. In order to understand the sense in which I intend these, it is instructive to consider the use of logics of knowledge and belief in AI and distributed computation. These logics, which were imported directly from analytic philosophy first to AI and then to other areas of computer science, describe the behavior of machines in terms of notions such as knowledge and belief. In computer science these mentalistic-sounding notions are actually given precise computational meanings, and are used not only to prove properties of distributed systems, but to program them as well. A typical rule in such a 'knowledge based' systems is "if processor A does not *know* that processor B has received its message, then processor A will not send the next message." AOP augments these logics with formal notions of choices, capabilities, commitments, and possibly others. A typical rule in the resulting systems will be "if agent A *believes* that agent B has *chosen* to do something harmful to agent A, then A will *request* that B change its choice." In addition, temporal information is included to anchor belief, choices and so on in particular points in time.

Here again we may benefit from some ideas in philosophy and linguistics. As in the case of knowledge, there exists work in exact philosophy on logics for choice and ability. Although they have not yet had an effect in AI comparable to that of logics of knowledge and belief, they may in the future.

---

[1] There is one more dimension to the comparison, which I omitted from the table, and it regards inheritance. Inheritance among objects is today one of the main features of OOP, constituting an attractive abstraction mechanism. I have not discussed it since it is not essential to the idea of OOP, and even less so to the idea of AOP. Nevertheless a parallel can be drawn here too.

Intentional terms such as knowledge and belief are used in a curious sense in the formal AI community. On the one hand, the definitions come nowhere close to capturing the full linguistic meanings. On the other hand, the intuitions about these formal notions do indeed derive from the everyday, common sense meaning of the words. What is curious is that, despite the disparity, the everyday intuition has proven a good guide to employing the formal notions in some circumscribed applications. AOP aims to strike a similar balance between computational utility and common sense.

2   Overview of the AOP Framework

A complete AOP system will include three primary components:

- A restricted formal language with clear syntax and semantics for describing a mental state. The mental state will be defined uniquely by several modalities, such as belief and commitment.
- An interpreted programming language in which to program agents, with primitive commands such as REQUEST and INFORM The semantics of the programming language will depend in part on the semantics of mental state.
- An 'agentifier,' converting neutral devices into programmable agents.

In the remainder of this paper I will start with an short discussion of the mental state. I will then present a general family of agent interpreters, a simple representative of which has already been implemented. I will end with a summary of recent research results and outstanding questions related to AOP.

3   On the Mental State of Agents

The first step in the enterprise is to define agents, that is, to define the various components of a mental state and the interactions between them. There is not a unique 'correct' definition, and different applications can be expected to call for specific mental properties.[2]

In related past research by others in AI three modalities were explored: belief, desire and intention (giving rise to the pun on BDI agent architectures). Other similar notions, such as goals and plans, were also pressed into service. These are clearly important notions; they are also complex ones, however, and not necessary the most primitive ones. Cohen and Levesque, for example, propose to reduce the notion of *intention* to those of *goal* and *persistence*. We too start with quite basic building blocks, in fact much more basic that those mentioned so far. We currently incorporate two modalities in the mental state of agents: *belief* and *obligation* (or *commitment*). We also define *decision* (or *choice*) as an obligation to oneself. Finally, we include a third cateogry which is not a mental

---

[2] In this respect our motivation here deviates from that of philosophers. However, I believe there exist sufficient similarities to make the connection between AI and philosophy mutually beneficial.

construct *per se*, *capability*. There is much to say on the formal definitions of these concepts; some of results described in the final section address this issue.

By restricting the components of mental state to these modalities I have in some informal sense excluded representation of motivation. Indeed, we do not assume that agents are 'rational' beyond assuming that their beliefs, obligations and capabilities are internally and mutually consistent. This stands in contrast to the other work mentioned above, which makes further assumptions about agents acting in their own best interests, and so on. Such stronger notions of rationality are obviously important, and I am convinced that in the future we will wish to add them. However, neither the concept of agenthood nor the utility of agent oriented programming depend on them.

4   A Generic Agent Interpreter

The behavior of agents is governed by programs; each agent is controlled by his own, private program. Agent programs themselves are not logical entities, but their control and data structures refer to the mental state of the agent using the logical language.[3]

The basic loop   The behavior of agents is, in principle, quite simple. Each agent iterates the following two steps at regular intervals:

1. Read the current messages, and update your mental state (including your beliefs and commitments);
2. Execute the commitments for the current time, possibly resulting in further belief change. Actions to which agents are committed include communicative ones such as informing and requesting.

The process is illustrated in Fig. 2; dashed arrows represent flow of data, solid arrows temporal sequencing.

5   Summary of Results and Ongoing Research

A more detailed discussion of AOP appears in (Shoham 1993); the implemented interpreter is documented in (Torrance 1991). Ongoing collaboration with the Hewlett Packard corporation is aimed at incorporating features of AOP in the New Wave™ architecture.

Work on mental state is proceeding on different fronts. Work on the 'statics' of mental state include (Moses and Shoham 1993), where we provide some results on the connection between knowledge and (one kind of) belief, and Thomas's (1992), which tackles the notions of capability, plan and intentions; more results on these topics are forthcoming. Other work begins to address dynamic aspects of

---

[3] However, an early design of agent programs by J. Akahani was entirely in the style of logic programming; in that framework program statements themselves were indeed logical sentences.

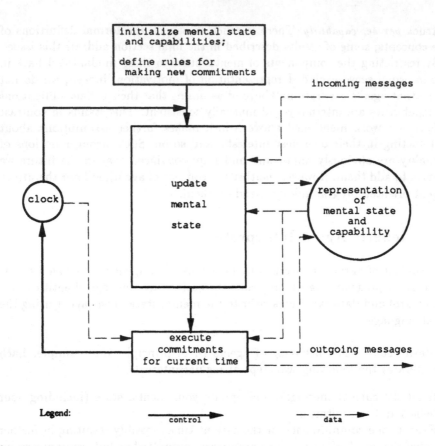

Fig. 2. A flow diagram of a generic agent interpreter

mental state. Lin and Shoham (1992b) investigate formal notions of memory and learning. Del Val and Shoham (to appear) address the logic of belief revision; specifically, the postulates of belief update are shown to be derivable from a formal theory of action. The theory used there is the 'provably correct' theory presented in (Lin and Shoham 1991).

In parallel to the logical aspects of action and mental state, we have investigated algorithmic questions. We have proposed a specific mechanism for tracking how beliefs change over time, called *temporal belief maps* (Isozaki and Shoham 1992). In (Brafman et al. 1993) we show that, similarly to distributed systems, the formal notion of knowledge can be applied to algorithmic robot motion planning.

Finally, we are interested in how multiple agents can function usefully in the presence of other agents. In particular, we are interested in mechanisms that minimize conflicts among agents, and have been investigating the utility of social laws in computational settings. Shoham and Tennenholtz (to appear) propose a general framework for representing social laws within a theory of action, and investigate the computational complexity of automatically synthesizing useful social laws; we also study a special case of traffic laws in a restricted robot environment. In (Shoham and Tennenholtz 1992) we study ways in which such conventions emerge automatically in a dynamic environment.

References

Brafman, R.I., J.-C. Latombe, and Y. Shoham: "Towards Knowledge-Level Analysis of Motion Planning," in: *Proc. AAAI*, Washington, 1993.

Del Val, A,. and Y. Shoham: "Belief update and theories of action," in: *Journal of Logic, Language and Information*, 1993 (to appear).

Dennett, D.C.: *The Intentional Stance*. Cambridge, MA: MIT Press, 1987.

Isozaki, H., and Y. Shoham: "A mechanism for reasoning about time and belief," in: *Proc. Conference on Fifth Generation Computer Systems*, Japan, 1992.

Lin, F., and Y. Shoham: "Provably correct theories of action," (preliminary report) in: *Proc. NCAI*, Anaheim, CA: 1991.

Lin, F., and Y. Shoham: *Concurrent actions in the situation calculus*. Stanford working document, 1992.

Lin, F., and Y. Shoham: *On the persistence of knowledge and ignorance*. Stanford working document, 1992.

McCarthy, J.: *Ascribing Mental Qualities to Machines*. Stanford AI Lab, Memo 326, 1979.

Moses, Y. and Y. Shoham: "Belief as Defeasible Knowledge," in: *Journal of Artificial Intelligence*, 64 (2) (1993), 299-322.

Shoham, Y.: "Agent Oriented Programming," in: *Journal of Artificial Intelligence*, 60 (1) (1993) 51-92.

Shoham, Y., and M. Tennenholtz: "Computational Social Systems: offline design," in: *Journal of Artificial Intelligence*, to appear.

Shoham, Y., and M. Tennenholtz: "Emergent conventions in multi-Agent systems," in: *Proc. KR*, Boston, 1992.

Thomas, B.: *A logic for representing action, belief, capability, and intention*. Stanford working document, 1992.

Torrance, M.: *The AGENT0 programming manual*. Stanford technical report, 1991.

An Application of Temporal Logic for Representation and Reasoning About Design

Keiichi Nakata*

Department of Artificial Intelligence, University of Edinburgh
80 South Bridge, Edinburgh EH1 1HN, Scotland
E-mail: keiichi@ai.ed.ac.uk

Abstract. This paper presents an approach to the application of temporal logic in order to represent and manipulate knowledge about physical systems in the design task. Focusing on the functional aspect of devices, we view the design task as achieving a desired behaviour of a device and its environment. Devices and environmental knowledge such as laws of physics are described in terms of causal rules, which are used to construct a model of the most likely behaviour of the system. Instead of giving a structural specification for design, a behavioural specification, which consists of the desired sequence of events, is provided, and it is compared with the predicted behaviour of the system for evaluation. The modification of the causal rule set is performed analogously to the design modification. Although computationally complex, this method sheds light on the formal treatment of temporal constraints and knowledge in design.

1 Introduction

One of the main features of an artefact is its function. The necessity for a certain function within the world is often the motivation to create an object, and an important factor which stimulates design. From this standpoint, it is reasonable to represent a device in terms of its functions. For instance, a door-bell has the function of notifying the people in the building when a visitor operates a trigger (typically a switch) outside. We can see that a function has its input and output, where input can be seen as the precondition to the consequent which is the output. If we apply a broad definition of 'causality', it is possible to describe functions as causations. In the case of a door-bell, we can see the operation of a switch as the 'cause' for the bell to ring.

Such representation and reasoning about design in terms of functions and behaviours can be found in the works by Freeman and Newell (1971) and Barrow (1984). The former suggested the validity of representing physical devices as functions and demonstrated the method of configuration task in terms of functions. The latter suggested a formal method of verifying the correctness of

* The author is funded by The Murata Overseas Scholarship Foundation and the Overseas Research Students Awards Scheme (Ref. ORS/9214048).

digital circuits by comparing the intended and actual behaviours. More recently, Sembugamoorthy and Chandrasekaran (1986) introduced *functional representation* as a formal representation for the functioning of a device at various levels of abstraction, and the work by Iwasaki and Chandrasekaran (1992) integrates functional representation with qualitative simulation for the design verification task. The work presented in this paper adds an additional aspect of description, namely a temporal feature. None of the contributions above provided an explicit way to represent temporal constraints.

The idea put forward in this paper is the application of temporal logic to represent and reason about the functions in design. Focusing on the characteristics of functionality of the devices as described above, the type of design problems we deal with here are those involving temporal transients, in other words, the design of artefacts which have significant temporal aspects. Also, we concentrate on the feasibility of the design from its functional point of view, and not optimisation. We represent and manage causal knowledge in the temporal logic devised by Yoav Shoham (1988). Although introduced as a means of dealing with the frame problems, the formality and the nonmonotonic feature of Shoham's logic suits nicely the description and the computation of the design task.

The main contribution of this paper is the investigation of the application of temporal logic to handle temporal constraints and maintain temporal relations between events in the design task. There has so far been a relatively small amount of research on the formal description and utilisation of temporal constraints in design. We are looking at practical, medium-size design problems and have so far worked on the domain of logic circuits. Primarily, we have tested on the configuration of set-reset flip-flop circuits, the verification of asynchronous circuits such as counters, and the configuration of JK flip-flop circuits. We obtained reasonable solutions for these problems, and based on this result we are evaluating the efficiency of the current implementation. The examples given in this paper are rather abstract to put emphasis on the logical foundation of this approach.

2 Formal Description of Design

The relationship between the world and an artefact can be formally described as follows:

$$E, S, D \models B \tag{1}$$

where

E: Environment (the laws of physics, causality)
S: Scenario (boundary conditions, sequence of factual events)
D: Design
B: Behaviour

133

This formula denotes that given an environment, a scenario and a design of an artefact, the behaviour of the system (including the environment) can be modelled. The design task is to create the set D, given E and S, which exhibits B, which is the desired behaviour. In order to construct D, all the others should be known, or at least be partially specified.

From now on we will use the word *system* to mean the union of the all premises (i.e., $E \cup S \cup D$).

Notice that this formulation suggests that the behavioural model B is non-monotonic. Suppose B_1 is a model such that

$$E, S, D \models B_1. \tag{2}$$

When a new condition p is added to the premise creating the model B_2 such that

$$(E, S, D) \wedge p \models B_2 \tag{3}$$

it is intuitively obvious that B_2 does not subsume B_1. For example, adding a condition that the device is to work under zero gravity to the existing design of a flushing toilet changes the behaviour of the system.

3 Shoham's Temporal Logic

Shoham (1988) introduces the language CI which deals with causality. The initial purpose of devising this language was to provide a temporal language which enables us to deal with frame problems, such as the Yale Shooting Problem. Logic CI is a nonmonotonic temporal language and provides the most likely model for a sound and consistent *causal theory* by chronologically minimising the changes. In other words, we prefer the model where things would remain the same as long as possible. In this way we can obtain a reasonable prediction given all the information available.

A causal theory Ψ consists of *causal rules* and *inertial rules*. A causal rule is of the form

$$\Phi \wedge \Theta \supset \Box\varphi \tag{4}$$

where

φ: a formula TRUE(t_1, t_2, p), indicating that the proposition p holds between time points t_1 and t_2

Φ: a (possibly empty) conjunction of the 'necessary' formulae $\Box\varphi$

Θ: a (possibly empty) conjunction of the 'contingent' formulae $\Diamond\varphi$

and describes that the conditions Φ and Θ *causes* $\Box\varphi$. It is important to note that the latest time point in the antecedent always precedes the initial time point of the consequent.[2]

Inertial rules have the same form as the causal rules but the consequents are potential terms POTEN$(t_1, t_2, p-q)$, indicating that the proposition q *potentially*

---

[2] Implying that the cause always precedes the effect.

holds, i.e., if nothing that prevents or terminates q is known, between time points t_1 and t_2.

Causal rules and inertial rules with empty Φ are called *boundary conditions* since their consequents are asserted as facts instead of causal consequence.

This language is attractive for our purpose, because:

- It is declarative.
- It provides a facility to represent default conditions.
- Given a sound and consistent causal theory, we can obtain a single model for the information available.
- It can be integrated with other techniques such as qualitative reasoning (Kuipers 1986).

Because it is declarative, we can readily add new causal rules and boundary conditions without altering existing rules. The 'contingent' terms can be regarded as default assumptions. Since there are many factors to be taken into consideration in design, it is often useful to be able to specify default conditions for efficiency.

A causal theory is *sound* when there are no conflicting defaults in the theory; it is *consistent* when no two rules with the same premise have contradictory conclusions. In the logic CI, a sound and consistent causal theory has a most preferred model, which is chronologically minimal.[3]

By extending the language from the propositional case to the first-order case (see (Bell 1991) for instance), we can incorporate other useful reasoning methods such as qualitative reasoning. This capability is advantageous when considering the integration of the reasoning techniques.

4 Behavioural Specification

Let us return to the formula 1. The design task was to construct D from other factors. E can be considered to be given since this is the knowledge about the world (in the sense of physics) and the domain specific knowledge. This leaves S and B to be provided. First we will focus on the latter, the behaviour of the system. We will begin by defining the *behavioural specification*.

Definition (Behavioural specification). The *behavioural specification* of design is the sequence of events and states which are desirable and necessary for the performance of the artefact. Formally, a behavioural specification BS is a set of atomic formulae φ_i with a temporal total ordering.

Intuitively, the behaviour of a system can be depicted by a sequence of events, which together with the events in the surrounding environment constitutes a *history*. The idea is to specify a history which is most desirable for the system. It is important to notice that it is not only the history of the behaviour of the

---

[3] *The unique c.m.i. theorem*, where c.m.i. stands for 'chronologically maximally ignorant'.

device, but of the whole system. We should take into account the changes in the environmental conditions that the operation of device might impose. For instance, the door-bell not only should ring, but should be noticed by those inside the building. The recognition of the bell ringing is beyond the behaviour of the door-bell itself, but is an important effect of the operation of the device.

The events specified in the behavioural specification should be totally ordered over time. We must specify the time point or an interval of time during which an event is taking place.

An example of a behavioural specification for the door-bell looks as follows.

$$BS = \{(1, 1, switch(on)), (2, 3, ring(bell)), (4, 4, notify(person(inside))))\}$$

Each event is represented as $(T_{begin}, T_{end}, Event)$ where T_{begin} and T_{end} respectively denote the initial and terminal time points in which the $Event$ takes place. The behavioural specification above then reads *"(we want something that when we) turn on the switch it rings the bell for an interval of time and notifies the person inside the building"*. For simplicity, appropriate integers are assigned to the time points which designates the total ordering of the events; however, this does not prevent us from assigning temporal symbols instead of integers and defining their orderings as we do later.

Now we move on to S, the scenario of the system. A scenario is a sequence of events which are known as facts. In causal theory, it is a set of boundary conditions, hence the events in a scenario are base formulae and potential terms.

A scenario in the door-bell example might be

$$S = \{\text{POTEN}(1, \infty, \text{p}-powerSupply(on)), \Box(1, \infty, deaf(person(inside))))\}$$

which reads, *"the power supply is assured to be on unless it is turned off, and the person inside the building is deaf"*.[4] The significance of having the set S separate from other factors is that unlike the environment, we can have more than one scenario for the prediction task. Assuming that there might be more than one possible behaviour of a device depending on the conditions we assume to be present, we can test them by merely replacing the scenario set without altering the other factors. This retains the modularity of knowledge involved.

To summarise, the factors in the system, the behavioural specification, scenario, environment and design have the following properties.

Behavioural specification is a set of desired events with temporal total order specifying their sequence $\varphi \in BS$. Given a behaviour B, ideally the purpose of design is to achieve $BS = B$.

Scenario is a set of boundary conditions which are the facts and the assumptions. $\Box\varphi$, POTEN$\varphi \in S$.

---

[4] The readers may notice that by giving a scenario that the person to be notified is deaf, the ringing a bell as has been specified in BS is not sufficient. This invokes the necessity of another device for the deaf person.

Environment is a set of physical laws and causal knowledge involved in the domain. $\Psi \in E$.

Design is a causal theory in which each rule represents a component of a device. $\Psi \in D$.

5 Design Verification

Once the prediction of the behaviour is obtained (we will call this the 'predicted behaviour'), the task now is to see if it fulfills the behavioural specification. Given a set of events in the behavioural specification and the predicted behaviour which is the set of events in the c.m.i. model, the design verification is to detect the difference between the two histories. If there is no discrepancy between the two, we can conclude that the current design with the provided scenario satisfies the specification. If discrepancies exist, then the design should be modified.

5.1 Detection of discrepancies

Discrepancies between two histories arise when the two have contradictory or different sequences of events. For instance, if the sequence of the events (propositions) p, q, r is actually $p \rightarrow q \rightarrow r$ in one history and $p \rightarrow r \rightarrow q$ in the other, the two histories can be considered to have a discrepancy. There is always a history that acts as a model (*behavioural specification*) and one that is to be verified (*predicted behaviour*).

We begin with identifying the basic discrepancies. As we discuss later, most discrepancies are expected to be a combination of the basic ones. Here I first describe the basic ones in the propositional case and discuss the first-order case.

5.2 Basic discrepancies

There are typically three basic discrepancies here: (1) insufficiency (propositions missing), (2) redundancy (have extra undesirable propositions), and (3) divergence (different histories). These cases are described in Table 1. In the first two cases, the histories compared are basically the same but with some parts being inconsistent. Precisely speaking, in the third there are two distinct cases: propositions q_n being irrelevant to p_n (which is equivalent to having '—' in their places) and q_n being $\neg p_n$. It should be noted that the time points (t_n) are not absolute, but relative: they merely describe the order in which these events occur.

These histories can be illustrated as described in Fig. 1. Notice that in the redundancy there might be a case when the extra proposition is not diverging (i.e., keeping the history on the same track). It should not affect the history for as long as it introduces no inconsistency.

MODEL	insufficient	redundant	divergence
(t_1,p_1)	(t_1,p_1)	(t_1,p_1)	(t_1,p_1)
(t_2,p_2)	(t_2,p_3)	(t_2,p_2)	(t_2,p_2)
(t_3,p_3)	(t_3,p_4)	(t_3,q)	(t_3,q_3)
(t_4,p_4)	(t_4,p_5)	(t_4,p_3)	(t_4,q_4)
(t_5,p_3)	—	(t_5,p_4)	(t_5,q_5)
—		(t_6,p_5)	—

Table 1. Three basic cases of discrepancies: *insufficient* — a proposition missing, *redundant* — an extra proposition, *divergence* — histories diverging.

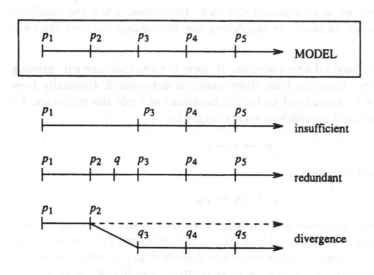

Fig. 1. Three cases of basic discrepancies.

In the first-order case, the basic discrepancies are the same as the propositional case except that the comparison of the events involves a unification process. For instance, if the next event occurring in the model is $value(1)$ and the prediction gives $value(X)$ (where X is a variable), then these two events are consistent provided that the variable X has not been instantiated to another value at the previous time point. Also owing to the semantics of the first-order logic, two terms are inconsistent not only when they are contradictory by negation (e.g., $p(a)$ and $\neg p(a)$) but semantically (e.g., $value(output(+))$ and $value(output(-))$).

5.3 Analysis of discrepancies

Comparison of the histories. As mentioned earlier, since the discrete time points attached to the events have no significance other than the definition of the ordering among the events, the sequence of events should be compared by their relative orderings. The basic strategy is to start from the earliest time points of two histories, and proceed chronologically until the first discrepancy is found. Since the behavioural specification consists of *necessary* sequence of events, all events in it should appear in the predicted behaviour. On the other hand, we can skip some of the intermediate events in the predicted history which do not appear in the behavioural specification as irrelevant or unimportant events. If the predicted behaviour contains all the events in the behavioural specification, we can judge that we have achieved the task. Otherwise, when the predicted behaviour misses out an event, we can detect the discrepancy between the two.[5]

Combination of basic discrepancies. If there is more than one rule missing, or redundant, more than one basic discrepancy is anticipated. Generally these discrepancies can be considered to be combinations of basic discrepancies. For example, the predicted sequence of events might be

$$p_1 \rightarrow q \rightarrow p_3$$

when the expected sequence is

$$p_1 \rightarrow p_2 \rightarrow p_3.$$

The predicted one can be seen as (1) missing the event p_2 and (2) having an extra event q. In this way, we attempt to break down various cases of discrepancies into basic ones. In the design modification stage described in the following section, each basic discrepancy can be dealt with separately, simplifying the task.

5.4 Compilation of devices

Once the behaviour of a device is modelled and proved to be plausible, it would be stored in the library for later use. For example, once the causal rule description of a solar cell is considered to be reasonable, it will be listed along with other devices. The rules can be made abstract if there is no secondary effect during its operation. Those events which would affect the environment (e.g., the increase in temperature, turning the light on, etc.) are called *external* events, as opposed to *internal* events, which occur only within a device (e.g., change in the flow within closed pipe circuit, increase of pressure within a container, etc.). Given an input event p_{input} and the output p_{output}, and the chains of events $q_1...q_n$ between them, i.e.,

$$p_{input} \rightarrow q_1 \rightarrow ... \rightarrow q_n \rightarrow p_{output},$$

---

[5] In cases where such intermediate events are not specified explicitly, we can assume that the specification 'doesn't care' how the succeeding events should be achieved.

we need only to keep explicit those external events among q_i which would affect other histories. In this causal chain, there are at most $n+1$ causal rules involved. Assuming that the external events are q_4 and q_{11}, then the set of rules can be reduced to

$$\Box(t, t, p_{input}) \wedge \Theta_1 \supset \Box(t + 1, t + 1, q_4)$$

$$\Box(t, t, q_4) \wedge \Theta_2 \supset \Box(t + 1, t + 1, q_{11})$$

$$\Box(t, t, q_{11}) \wedge \Theta_3 \supset \Box(t + 1, t + 1, p_{output}).$$

These rules will be the compiled library of the device. It would preserve the temporal order of events, and eliminate events which are irrelevant to the overall behaviour of the device. The compiled library is labelled with the domain in which it was created and loaded when an appropriate environment was specified. This acts as an assignment of the resource available.

6 Design Modification

As the result of design verification, we can identify the point where modifications are necessary in order to obtain a desirable behaviour of the system. Since the primary purpose at this stage is to obtain the event which is essential to occur in order to meet the behavioural specification, the strategy is to assert a rule which 'causes' the desired event. The rule asserted here is typically very abstract. We then proceed to 'prove' the rule using existing rules in the system. The idea is to bring the cause and the effect of the newly added rule as close as possible. Finally, we attempt to bridge the gap by devising a component which provides the effect of the causal rule obtained after this proof procedure.

6.1 Rule assertion and its proof procedure

First we begin by asserting a causal rule which would give the desirable history for the current design. This rule is a very abstract rule which may have no intuitive causality whatsoever, but has an effect of 'forcing' an event which is desirable to occur at a time point later in the occurrence of the premise given. Being a causal rule, it would look like

$$\Box \varphi_P \wedge \Box \varphi_T \supset \Box \varphi_Q \tag{5}$$

where φ_P and φ_Q are of the form $\text{TRUE}(t_1, t_2, p)$ (base formula, p is a proposition or a first-order formula) and φ_T is a formula that describes the ordering of the temporal variables in the rule.

The object of the operation is to rewrite the causal rule (5) and prove it by existing causal rules. In order to achieve this, the following operations are applied to the antecedent and the consequent of the rule.

- *Antecedent*: Find an alternative rule
- *Consequent*: Perform abduction

Essentially, for $\Box\varphi_P$ in the antecedent, find a causal rule that contains this formula in its antecedent. If such rule exists, say $\Phi_{P1} \wedge \Theta_{P1} \supset \Box\varphi_{P1}$, then $\Box\varphi_P \in \Phi_{P1}$. For the consequent, find a causal rule that has $\Box\varphi_Q$ as its consequent. Assume that such a rule exists, say $\Phi_{Q1} \wedge \Theta_{Q1} \supset \Box\varphi_Q$. Create a rule of the form

$$\Box\varphi_{P1} \wedge \Box\varphi_{T1} \supset \Box\varphi_{Q1} \tag{6}$$

where $\Box\varphi_{Q1} \in \Phi_{Q1}$. Be sure that the temporal constraint $\Box\varphi_T$ and $\Box\varphi_{T1}$ are consistent. A simple example of these operations is illustrated in Fig. 2.

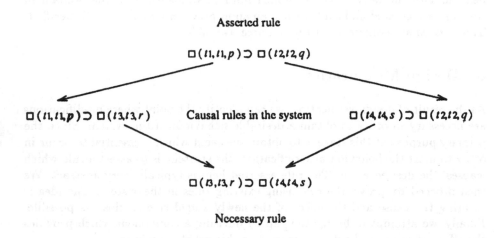

Fig. 2. Proof operation of an asserted rule.

At this point there are three possible cases:

1. Rule (6) resembles the existing rule
2. Further operations can be applied
3. Dead end

The causal rule constructed in this manner does not result in the reconstruction of an existing rule, since if there were such rule in the causal theory, it would have been applied at the modelling stage. In the first case, the constructed rule would look like an existing rule but does not subsume it. For instance, a necessary condition could be missing in the new rule, but with all other conditions in the antecedent and the consequent being the same. In this case, rather than just asserting such rule, it would be more reasonable to make the existing rule being applied by devising the missing condition to be fulfilled. Another possible case of resemblance is the difference in the default conditions, but this could end up violating the soundness condition. The second case in the list suggests that the system may be decomposed further. Otherwise (the third case) nothing can be done to it, which suggest that there is no such device present in the current design, or "a device which has such function is necessary". By asserting this rule, we modify the design D and can achieve the required behaviour.

This procedure so far gives the partial solution to the problem since the following must be proved.

- The validity of $\Box \varphi_i \in \Phi_{P1}$ $(i \neq P)$
- The validity of $\Box \varphi_j \in \Phi_{Q1}$ $(j \neq Q)$

That is, the premises other than the one we used in order to construct the new rule must be justified to make this assertion of the new rule valid. In reality, the structure of causal rules are not as simple as those described in Fig. 2. There are typically more than one conditions in the premise, and there may be some default conditions. In such cases, each condition in the premise should be proved by the causal rules in the system, which invokes the chain of proof operations.

The proof is by repetitive application of operations to bring together the antecedent and the consequent of the necessary rule as close as possible.

6.2 Preserving temporal constraints

The constraint for the temporal terms are represented in one of the cases below:

1. Temporal constants ($\in I$) used within base terms: in this case numerical ordering corresponds to the temporal order
2. Temporal variables:
   - Differential description — $t + m \leq t + n$ if $m \leq n$ $m, n \in I$, used within base terms
   - Constraining term — used outside base terms which is of the form $\Box(t_1 \leq ... \leq t_n)$ (or $\Diamond(t_1 \leq ... \leq t_n)$) where t_i are temporal terms appearing in the base term
   - A combination of the two

Temporal constants are typically assigned initially to the boundary conditions. A differential description is typically used for the potential histories (to set up simulation environments). Since the explicit description of the constraining terms are preferred for causal and interval rules, the preservation of temporal constraints in these terms is crucial.

Temporal constraint preservation in rule synthesis. In any operation over an antecedent or the consequent (i.e., expansion using existing rules) the local temporal constraints apply, which preserves the temporal ordering between the temporal variables. This applies to both the causal and interval rules. It is necessary, however, to maintain the temporal ordering in the rule synthesis.

Let $\Box\varphi_A$ and $\Box\varphi_B$ be the antecedent and the consequent of the new (synthesised) rule, which looks like

$$\Box\varphi_A \supset \Box\varphi_B \tag{7}$$

and $t_A = ltp(\varphi_A)$ and $t_B = ltp(\varphi_B)$, where $ltp(\varphi)$ denotes the latest time point for the base formula φ. The syntactic constraint then will be $t_A < t_B$, which should be added as the 'necessary' condition,[6]

$$\Box\varphi_A \wedge \Box(t_A < t_B) \supset \Box\varphi_B \tag{8}$$

6.3 Removing components

Since each rule in the causal theory for design D represents a function of a component, asserting a rule to the set is analogous to adding a new component to the configuration of design. Similarly, the removal or the replacement of a component are performed by removing one or more rules from the theory.

Retracting a causal rule is equivalent to removing a function from which an undesirable event was derived, and, as a result, detaching a component which was responsible for the behaviour. There are essentially two cases for this.

- The causal rule which concluded the event is retractable (i.e., design component rules).
- The causal rule which concluded the event is *not* retractable (i.e., environment causal rules such as laws of physics).

In the first case, removing the rule solves the problem.[7] It is equivalent to removing a component in the design. An alternative is to keep the rule but suppress it from firing by removing the events which satisfy the premise. We will follow the first strategy and not the second since it is always possible to reconstruct the rules which would effectively have the same result.

In the second case, since it is not reasonable to remove the rule (unless we can alter the environment), the strategy will be the second of those mentioned in

---

[6] There is an implicit assumption that there is a well-defined temporal ordering between the temporal variables within a base formula; the *ltps* become undecidable otherwise.

[7] The implication of removing of the rule can be more problematic. In some cases, the rule may have concluded a desirable event elsewhere, which no longer could be supported. It is, however, possible to reconstruct a causal rule which concludes the desirable event but with different premises: this will invoke yet another design problem.

the previous paragraph, to suppress the rules from firing. We first take a look at the causal rule and see what were the events that caused the rule to fire. The aim now is to choose an event among them and prevent it. By checking which causal rule caused the event, we would have the same task we originally had, but with different events and rules, and at preceding time point. This process continues until we find a retractable rule, or possibly a boundary condition. Theoretically, this process should terminate at some point. The soundness and the consistency of the causal theory must be maintained by monitoring the changes in the rules.

7 Discussion

7.1 The validity of c.m.i. model as the prediction

The unique c.m.i. model shows that there is only one most preferable model for a causal theory (Shoham 1988). This might be too restrictive if we consider the possibility of having multiple possible behaviours of a design. To focus on the most likely outcome would raise the danger of overlooking a non-deterministic aspect of the design. It might be more informative if we could maintain all possible behaviours in order to evaluate the design.

Focusing on the unique model, however, does not exclude other possible behaviours. It is merely suggested as 'the most likely behaviour' under 'given conditions and (default) assumptions'. Typically, a causal theory would result in having different histories depending on whether the default is overridden, since there is no need to state the default conditions in a causal rule when they do not affect the causal outcomes. Whenever the predicted behaviour contradicts with the expected behaviour, there is always a possibility that the default assumed is not appropriate. We can check this by listing out the default conditions used in the modelling process, and if there is any 'wrong' default, it is easily overridden by adding a new boundary condition.[8] In most cases, multiple worlds occur due to lack of information about specific conditions that distinguish the worlds.

There is another case where there arises more than one behaviour. When a variable is universally quantified and there is no need to assign any value to it, there are as many possible worlds as the number of members in the set of individuals semantically assigned to the variable. A typical example of this case is the 'don't care' state in the logic gates. Since there is no restrictions to the value of the variable, assuming the 'most likely' value to it seems to be plausible.

The reason why we think a unique model is appropriate is that it indicates how a device would behave by default. If the designer is satisfied with the outcome, using the default might be good enough, and if not, s/he can find out which default to override. In case of multiple worlds, we can see what 'can' happen, but sometimes the number of alternatives is too high to manage. The process of choosing the 'right' alternative one would nevertheless be the same.[9]

---

[8] By definition, boundary conditions are □ ('necessary') events.

[9] In this context, we are assuming that a device would behave uniquely under given condition. There are cases when a device should have multiple behaviour, but we think this happens only when the design is under-constrained.

7.2 Unachievable specification

The evaluation-modification cycle of the design process does not always work, especially when there is a conflict in the initial specification. Since one of the results to be obtained after the proof is the construction of a new causal rule (whose causality is necessary to achieve the specification), the new causal rule can be reasonable one or not. For example, the causal rule might give a description of a device which reads "when the power of the magnetic field increases, the mass decreases", or "when a given object is travelling at the speed of light, it accelerates."[10] When such impossible causation is derived as the most plausible among other alternatives, there is something wrong with the behavioural specification.

In such case, there is no way out except to modify the specification. Since there will be some index to the boundary conditions that are responsible for the undesirable events after the failure of the proof, it is possible to figure out the conditions under conflict. Automatic resolution of these conflicts is not in the scope of the current research, but it should be possible to support the designer by indicating the necessary conditions to obtain the desired behaviour.

8 Conclusion

The essence of the approach illustrated in this paper is to treat the functionality of design components in terms of causality. Causal relations among events produce a chronological sequence of events (or a history), which is the behaviour of the whole system in the given environment, scenario and the causal rules describing the design components. We assign the specification of the device in terms of such behaviour, and the design task is to create a causal theory which would have such behaviour as its most likely outcome. The causal rules were manipulated according to Shoham's temporal logic which incorporates the nonmonotonic aspect of the prediction task. The formulation is based on a well-defined theory, with clearly defined theoretical justification.

The advantages of this approach are:

- the behavioural specification of devices allows more intuitive descriptions of the functionality of the system, especially in dynamic systems where the devices interact with the environment, and
- it utilises temporal constraints for design evaluation and modification, which enables the representation of delays and relative timings in the occurrence of the events. Without temporal logic, we must rely on the implicit ordering of events.

On the other hand, the main drawback in the design stage of this approach is its complexity. There are typically a number of rule candidates for the abduction

---

[10] The former is the case when the causal connection is not obvious or non-empirical, but *may be* possible with additional knowledge about physics. The latter is simply impossible to have such an effect in real physics.

process, and the combinatorics become more complex as the number of such rules and the number of conditions in their premise become greater. At this moment, some domain specific heuristics are used to reduce the search space, and some choices are reserved for the human user to decide. This permits the users to incorporate their preference, but as the number of rules increase in more complex domain, it is necessary to introduce more elaborate optimisation procedures to guide search. We are currently investigating the construction of 'most likely' models backwards in time to guide the abduction process. This would limit the number of plausible choices and provide suggestions based on some preference criteria.

Acknowledgements

I would like to thank Dave Robertson and Alan Smaill for fruitful discussions and useful comments on earlier drafts of this paper.

References

Barrow, H. G.:"VERIFY: A Program for Proving Correctness of Digital Hardware Designs," in: *Artificial Intelligence* 24 (1984) 437–491.

Bell, J.: "Extended causal theories," in: *Artificial Intelligence* 48 (1991) 211–224.

Freeman, P., and A. Newell: "A model for functional reasoning in design," in: *Proc. IJCAI-71* (1971) 621–640.

Iwasaki, Y.,and B. Chandrasekaran: "Design verification through function- and behavior- oriented representations: bridging the gap between function and behavior," in: J.S. Gero (ed.), *Artificial Intelligence in Design '92*, Pittsburgh, pages 597–616. Dordrecht: Kluwer Academic Publishers, 1992.

Kuipers, B.: "Qualitative Simulation," in: *Artificial Intelligence* 29 (1986) 289–338.

Sembugamoorthy, V., and B. Chandrasekaran: "Functional Representation of Devices and Compilation of Diagnostic Problem-Solving Systems," in: J.L. Kolodner and C.K. Riesbeck (eds), *Experience, Memory and Reasoning*, pages 47–73. Hillsdale, NJ: Lawrence Erlbaum Associates, 1986.

Shoham, Y.: *Reasoning about Change*. Cambridge, MA: The MIT Press, 1988.

Knowledge Theoretic Properties
of Topological Spaces

Konstantinos Georgatos

Department of Mathematics
Graduate School and University Center
City University of New York
33 West 42nd Street
New York, NY 10036
E-mail: koghc@cunyvm.cuny.edu

Abstract. We study the topological models of a logic of knowledge
for topological reasoning, introduced by Larry Moss and Rohit Parikh
(1992). Among our results is the confirmation of a conjecture by Moss
and Parikh, as well as the finite satisfiability property and decidability
for the theory of topological models.

1 Introduction

We are unable to measure natural quantities with exact precision. Physical devices or bounded resources always introduce a certain amount of error. Because
of this fact, we are obliged to limit our observations to approximate values or,
better, to sets of possible values. Whenever this happens, sets of points, rather
than points, are our subject of reasoning. Thus, the statement *"the amount of
ozone in the upper atmosphere has decreased by 12 per cent"* can never be known
to be true with this precision. What we mean is that the decrease has a value
in the interval $(12 - \epsilon, 12 + \epsilon)$ for some positive real number ϵ. If we are able
to spend more resources (taking more samples, using more precise instruments,
etc.) we may be able to affirm that the value belongs to a smaller interval and
therefore to refine our observation. The topology of intervals in the real line is
our domain of reasoning.

The above limitations do not prevent us from drawing conclusions. In fact
it is enough that we *know* that a certain quantity belongs to a set of possible values. The first hundred decimal points of π are enough for most practical purposes and if we decide to settle for such a value, an algorithm that
computes these decimal points conveys the same knowledge as the actual algorithm that computes π. What we know is exactly the common properties of
all algorithms belonging to the same open set of the algorithms we observe in
the topology of initial segments and this notion of knowledge coincides with
the traditional one (Hintikka 1962, Fagin et al. 1991, Halpern and Moses 1984,
Parikh and Ramanujam 1985): what is known is whatever is true in all states
compatible with the observer's view.

Increase of knowledge is strongly linked with the amount of resources we are
willing to invest. An increase of information is accompanied by an increase in the

effort of acquiring it. This corresponds to the refinement of the open set in the relevant topology. In the formal system introduced in (Moss and Parikh 1992) there are two basic modal operators; K for *knowledge* and □ for *effort*.

A basic advantage of this logic and its semantics over other temporal logics or logics of change is that, though we make no mention of set, we are able to interpret assertions relative to a set of possible states and, at the same time, keep the dependence on the actual state. Topology is a tool for modelling classes of statements with an intuitionistic flavor such as refutative or affirmative assertions (see (Vickers 1989)) and Moss and Parikh's system enables us to treat them in a classical modal framework. In many respects the way we interpret the modal operator K resembles the algebraic semantics of a modal operator used to interpret intuitionistic statements as in (Rasiowa and Sikorski 1968). As the intuitionistic version of a statement is the interior of the subset that represents the classical version of it, KA is satisfied only in the open subsets which are subsets of the interior of the set of points which satisfy a property A.

The fundamental reasoning that this logic tries to capture has many equivalents in recursion theory and elsewhere in Mathematics. The discussion of them is well beyond the scope of this paper and the reader is referred to (Georgatos 1993) and (Moss and Parikh 1992) for a more detailed exposition.

In the following section, we describe the syntax and semantics of the logic and we give complete axiomatisations with respect to subset spaces and topological spaces. In Section 3 we develop a theory for describing the validity problem in topological spaces. In Section 4 we study the model based on the basis of a topological space closed under finite unions, and we prove it equivalent to the topological space that it generates. These results translate to a completeness theorem for topologies, given a finite axiomatisation for the class of spaces which are closed under (finite) intersection and union. In the last section we prove finite satisfiability for the class of topological models and decidability for their theory.

2 The logic

We follow the notation of (Moss and Parikh 1992).

Our language is bimodal and propositional. Formally, we start with a countable set A of *atomic formulae* containing two distinguished elements ⊤ and ⊥. Then the *language* $\mathcal{L}$ is the least set such that A $\subseteq \mathcal{L}$ and $\mathcal{L}$ is closed under the following rules:

$$\frac{\phi, \psi \in \mathcal{L}}{\phi \wedge \psi \in \mathcal{L}} \qquad \frac{\phi \in \mathcal{L}}{\neg \phi, \Box \phi, \mathsf{K}\phi \in \mathcal{L}}$$

The above language can be interpreted inside any spatial context.

Definition 1. Let X be a set and $\mathcal{O}$ a subset of the powerset of X, i.e., $\mathcal{O} \subseteq \mathcal{P}(X)$ such that $X \in \mathcal{O}$. We call the pair $\langle X, \mathcal{O} \rangle$ a *subset space*. A *model* is a triple $\langle X, \mathcal{O}, i \rangle$, where $\langle X, \mathcal{O} \rangle$ is a subset space and i a map from A to $\mathcal{P}(X)$ with $i(\top) = X$ and $i(\bot) = \emptyset$ called *initial interpretation*.

We denote the set $\{(x, U) \mid U \in \mathcal{O}, \ x \in U\} \subseteq X \times \mathcal{O}$ with $X \dot{\times} \mathcal{O}$. For each $U \in \mathcal{O}$ let $\downarrow U$ be the set $\{V \mid V \in \mathcal{O}$ and $V \subseteq U\}$, i.e., the lower closed set generated by U in the partial order $(\mathcal{O}, \subseteq)$.

Definition 2. The *satisfaction relation* $\models_{\mathcal{M}}$, where $\mathcal{M}$ is the model $\langle X, \mathcal{O}, i \rangle$, is a subset of $(X \dot{\times} \mathcal{O}) \times \mathcal{L}$ defined recursively by (we write $x, U \models_{\mathcal{M}} \phi$ instead of $((x, U), \phi) \in \models_{\mathcal{M}})$:

$$x, U \models_{\mathcal{M}} A \quad \text{iff} \quad x \in i(A), \text{ where } A \in \mathbf{A}$$

$$x, U \models_{\mathcal{M}} \phi \wedge \psi \ \text{if} \ \ x, U \models_{\mathcal{M}} \phi \ \text{and} \ x, U \models_{\mathcal{M}} \psi$$

$$x, U \models_{\mathcal{M}} \neg \phi \quad \text{if} \quad x, U \not\models_{\mathcal{M}} \phi$$

$$x, U \models_{\mathcal{M}} K\phi \quad \text{if} \quad \text{for all } y \in U, \ \ y, U \models_{\mathcal{M}} \phi$$

$$x, U \models_{\mathcal{M}} \Box \phi \quad \text{if} \quad \text{for all } V \in \downarrow U \text{ such that } x \in V, \ \ x, V \models_{\mathcal{M}} \phi.$$

If $x, U \models_{\mathcal{M}} \phi$ for all (x, U) belonging to $X \dot{\times} \mathcal{O}$ then ϕ is *valid* in $\mathcal{M}$, denoted by $\mathcal{M} \models \phi$.

We abbreviate $\neg \Box \neg \phi$ and $\neg K \neg \phi$ with $\Diamond \phi$ and $L\phi$ respectively. We have that

$$x, U \models_{\mathcal{M}} L\phi \ \text{if there exists } y \in U \text{ such that } y, U \models_{\mathcal{M}} \phi$$

$$x, U \models_{\mathcal{M}} \Diamond \phi \ \text{if there exists } V \in \mathcal{O} \text{ such that } V \subseteq U, \ x \in V, \text{ and } x, V \models_{\mathcal{M}} \phi.$$

Many topological properties are expressible in this logical system in a natural way. For instance, in a model where the subset space is a topological space, $i(A)$ is *open* whenever $A \to \Diamond KA$ is valid in this model or $i(A)$ is *nowhere dense* whenever $L \Diamond K \neg A$ is valid (cf. (Moss and Parikh 1992)).

Example 1. Consider the set of *real numbers* $\mathbf{R}$ with the usual topology of open intervals. We define the following three predicates:

$$\mathbf{pi} \text{ where } i(\mathbf{pi}) = \{\pi\}$$
$$\mathbf{I}_1 \text{ where } i(\mathbf{I}_1) = (-\infty, \pi]$$
$$\mathbf{I}_2 \text{ where } i(\mathbf{I}_2) = (\pi, +\infty)$$
$$\mathbf{Q} \text{ where } i(\mathbf{Q}) = \{q \mid q \text{ is rational }\}.$$

There is no real number p and open set U such that $p, U \models K\mathbf{pi}$ because that would imply $p = \pi$ and $U = \{\pi\}$ and there are no singletons which are open.

A point x belongs to the *closure* of a set W if every open U that contains $x \cap W$. Thus π belongs to the closure of $(\pi, +\infty)$, i.e., every open U that contains π has a point in $(\pi, +\infty)$. This means that for all U such that $\pi \in U$, $\pi, U \models L\mathbf{I}_2$, therefore $\pi, \mathbf{R} \models \Box L\mathbf{I}_2$. Following the same reasoning $\pi, \mathbf{R} \models \Box L\mathbf{I}_1$, since π belongs to the closure of $(-\infty, \pi]$.

A point x belongs to the *boundary* of a set W whenever x belong to the closure of W and $X - W$. By the above, π belongs to the boundary of $(-\infty, \pi]$ Among our results is a solution of a conjecture by the form and $\pi, \mathbf{R} \models \Box(L\mathbf{I}_1 \wedge L\mathbf{I}_2)$.

A set W is *closed* if it contains its closure. The interval $i(I_1) = (-\infty, \pi]$ is closed and this means that the formula $\Box L I_1 \rightarrow I_1$ is valid.

A set W is *dense* if all opens contain a point of W. The set of rational numbers is dense which translates to the fact that the formula $\Box L Q$ is valid. To exhibit the reasoning in this logic, suppose that the set of rational numAmong our results is a solution of a conjecture by the formbers was closed then both $\Box L Q$ and $\Box L Q \rightarrow Q$ would be valid. This implies that Q would be valid which means that all reals would be rationals. Hence the set of rational numbers is not closed.

The following set of axioms and rules, denoted by **MP***, is sound and complete for the class of topological spaces (see (Georgatos 1993)) while axioms 1 through 10, denoted by **MP**, appeared first and proven sound and complete for the class of subset spaces in (Moss and Parikh 1992). Among our results is a solution ofAmong our results is a solution of a conjecture by the form a conjecture by the form

Axioms

1. All propositional tautologies
2. $(A \rightarrow \Box A) \land (\neg A \rightarrow \Box \neg A)$, for $A \in \mathsf{A}$
3. $\Box(\phi \rightarrow \psi) \rightarrow (\Box\phi \rightarrow \Box\psi)$
4. $\Box\phi \rightarrow \phi$
5. $\Box\phi \rightarrow \Box\Box\phi$
6. $\mathsf{K}(\phi \rightarrow \psi) \rightarrow (\mathsf{K}\phi \rightarrow \mathsf{K}\psi)$
7. $\mathsf{K}\phi \rightarrow \phi$
8. $\mathsf{K}\phi \rightarrow \mathsf{K}\mathsf{K}\phi$
9. $\phi \rightarrow \mathsf{K}\mathsf{L}\phi$
10. $\mathsf{K}\Box\phi \rightarrow \Box\mathsf{K}\phi$
11. $\Diamond\Box\phi \rightarrow \Box\Diamond\phi$
12. $\Diamond(\mathsf{K}\phi \land \psi) \land \mathsf{L}\Diamond(\mathsf{K}\phi \land \chi) \rightarrow \Diamond(\mathsf{K}\Diamond\phi \land \Diamond\psi \land \mathsf{L}\Diamond\chi)$

Rules

$$\frac{\phi \rightarrow \psi, \phi}{\psi} \; \text{MP}$$

$$\frac{\phi}{\mathsf{K}\phi} \; \text{K-Necessitation} \qquad \frac{\phi}{\Box\phi} \; \Box\text{-Necessitation}$$

3 Stability and Splittings

Suppose that X is a set and $\mathcal{T}$ a topology on X. In the following we assume that we are working in the topological space $(X, \mathcal{T})$. Our aim is to find a partition of $\mathcal{T}$, where a given formula ϕ "retains its truth value" for each point throughout a member of this partition. We shall show that there exists a finite partition of this kind.

Definition 3. Given a finite family $\mathcal{F} = \{U_1, \ldots, U_n\}$ of opens, we define the *remainder* of (the principal ideal in $(\mathcal{T}, \subseteq)$ generated by) U_k by

$$\mathsf{Rem}^{\mathcal{F}} U_k \;=\; \downarrow U_k - \bigcup_{U_k \not\subseteq U_i} \downarrow U_i.$$

Proposition 4. *In a finite set of opens* $\mathcal{F} = \{U_1, \ldots, U_n\}$ *closed under intersection, we have*

$$\mathsf{Rem}^{\mathcal{F}} U_i \;=\; \downarrow U_i - \bigcup_{U_j \subset U_i} \downarrow U_j,$$

for $i = 1, \ldots, n$.

Proof.
$$
\begin{aligned}
\mathsf{Rem}^{\mathcal{F}} U_i &= \downarrow U_i - \bigcup_{U_i \not\subseteq U_h} \downarrow U_h \\
&= \downarrow U_i - \bigcup_{U_i \not\subseteq U_h} \downarrow (U_h \cap U_i) \\
&= \downarrow U_i - \bigcup_{U_j \subset U_i} \downarrow U_i.
\end{aligned}
$$

We denote $\bigcup_{U_i \in \mathcal{F}} \downarrow U_i$ with $\downarrow \mathcal{F}$.

Proposition 5. *If* $\mathcal{F} = \{U_1, \ldots, U_n\}$ *is a finite family of opens, closed under intersection, then*

1. $\mathsf{Rem}^{\mathcal{F}} U_i \cap \mathsf{Rem}^{\mathcal{F}} U_j = \emptyset$, *for* $i \neq j$,
2. $\bigcup_{i=1}^{n} \mathsf{Rem}^{\mathcal{F}} U_i = \downarrow \mathcal{F}$, *i.e.,* $\{\mathsf{Rem}^{\mathcal{F}} U_i\}_{i=1}^{n}$ *is a partition of* $\downarrow \mathcal{F}$. *We call such an* $\mathcal{F}$ *a finite splitting (of* $\downarrow \mathcal{F}$),
3. *if* $V_1, V_3 \in \mathsf{Rem}^{\mathcal{F}} U_i$ *and* V_2 *is an open such that* $V_1 \subseteq V_2 \subseteq V_3$ *then* $V_2 \in \mathsf{Rem}^{\mathcal{F}} U_i$, *i.e.,* $\mathsf{Rem}^{\mathcal{F}} U_i$ *is convex.*

Proof. The first and the third are immediate from the definition.
For the second, suppose that $V \in \downarrow \mathcal{F}$ then $V \in \mathsf{Rem}^{\mathcal{F}} \bigcap_{V \in \downarrow U_i} U_i$.

Every partition of a set induces an equivalence relation on this set. The members of the partition comprise the equivalence classes. Since a splitting induces a partition, we denote the equivalence relation induced by a splitting $\mathcal{F}$ by $\sim_{\mathcal{F}}$.

Definition 6. Given a set of open subsets $\mathcal{G}$, we define the relation $\sim'_{\mathcal{G}}$ on $\mathcal{T}$ with $V_1 \sim'_{\mathcal{G}} V_2$ if and only if $V_1 \subseteq U \Leftrightarrow V_2 \subseteq U$ for all $U \in \mathcal{G}$.

We have the following

Proposition 7. *The relation* $\sim'_{\mathcal{G}}$ *is an equivalence.*

Proposition 8. *Given a finite splitting* $\mathcal{F}$, $\sim'_{\mathcal{F}} = \sim_{\mathcal{F}}$ *i.e., the remainders of* $\mathcal{F}$ *are the equivalence classes of* $\sim'_{\mathcal{F}}$.

Proof. Suppose $V_1 \sim'_{\mathcal{F}} V_2$ then $V_1, V_2 \in \mathsf{Rem}^{\mathcal{F}} U$, where

$$U = \bigcap \{\, U' \mid V_1, V_2 \subseteq U,\ U' \in \mathcal{F} \,\}.$$

For the other way suppose $V_1, V_2 \in \mathsf{Rem}^{\mathcal{F}} U$ and that there exists $U' \in \mathcal{F}$ such that $V_1 \subseteq U'$ while $V_2 \nsubseteq U'$. Then we have that $V_1 \subseteq U' \cap U$, $U' \cap U \in \mathcal{F}$ and $U' \cap U \subseteq U$ i.e., $V_1 \notin \mathsf{Rem}^{\mathcal{F}} U$.

We state some useful facts about splittings.

Proposition 9. *If $\mathcal{G}$ is a finite set of opens, then $\mathsf{Cl}(\mathcal{G})$, its closure under intersection, yields a finite splitting for $\downarrow\mathcal{G}$.*

The last proposition enables us to give yet another characterization of remainders: every family of points in a complete lattice closed under arbitrary joins comprises a *closure system*, i.e., a set of fixed points of a closure operator of the lattice (cf. (Gierz et al. 1980).) Here, the lattice is the poset of the opens of the topological space. If we restrict ourselves to a finite number of fixed points then we just ask for a finite set of opens closed under intersection i.e., Proposition 9. Thus a closure operator in the lattice of the open subsets of a topological space induces an equivalence relation, two opens being equivalent if they have the same closure, and the equivalence classes of this relation are just the remainders of the open subsets which are fixed points of the closure operator. The maximum open in $\mathsf{Rem}^{\mathcal{F}} U$, i.e., U, can be taken as the representative of the equivalence class which is the union of all open sets belonging to $\mathsf{Rem}^{\mathcal{F}} U$.

We now introduce the notion of stability corresponding to what we mean by "a formula retains its truth value on a set of opens".

Definition 10. If $\mathcal{G}$ is a set of opens then $\mathcal{G}$ is *stable for* ϕ, if for all x, either $x, V \models \phi$ for all $V \in \mathcal{G}$, or $x, V \models \neg\phi$ for all $V \in \mathcal{G}$, such that $x \in V$.

Proposition 11. *If $\mathcal{G}_1, \mathcal{G}_2$ are sets of opens then*

1. *if $\mathcal{G}_1 \subseteq \mathcal{G}_2$ and $\mathcal{G}_2$ is stable for ϕ then $\mathcal{G}_1$ is stable for ϕ ,*
2. *if $\mathcal{G}_1$ is stable for ϕ and $\mathcal{G}$ is stable for χ then $\mathcal{G}_1 \cap \mathcal{G}_2$ is stable for $\phi \wedge \chi$.*

Proof. (a) is easy to see, while (b) is a corollary of (a).

Definition 12. A finite splitting $\mathcal{F} = \{U_1, \ldots, U_n\}$ is called a *stable splitting for* ϕ, if $\mathsf{Rem}^{\mathcal{F}} U_i$ is stable for ϕ for all $U_i \in \mathcal{F}$.

Proposition 13. *If $\mathcal{F} = \{U_1, \ldots, U_n\}$ is a stable splitting for ϕ, so is*

$$\mathcal{F}' = \mathsf{Cl}(\{U_0, U_1, \ldots, U_n\}),$$

where $U_0 \in \downarrow\mathcal{F}$.

Proof. Let $V \in \mathcal{F}'$ then there exists $U_l \in \mathcal{F}$ such that $\mathsf{Rem}^{\mathcal{F}'} V \subseteq \mathsf{Rem}^{\mathcal{F}} U_l$ (e.g., $U_l = \bigcap\{U_i | U_i \in \mathcal{F}, V \subseteq U_i\}$) i.e., $\mathcal{F}'$ is a *refinement* of $\mathcal{F}$. But $\mathsf{Rem}^{\mathcal{F}} U_l$ is stable for ϕ and so is $\mathsf{Rem}^{\mathcal{F}'} V$ by Proposition 11(a).

The above proposition tells us that if there is a finite stable splitting for a topology then there is a closure operator with finitely many fixed points whose associated equivalence classes are stable sets of open subsets.

Suppose that $M = \langle X, T, i \rangle$ is a topological model for $\mathcal{L}$. Let $\mathcal{F}_M$ be a family of subsets of X generated as follows: $i(A) \in \mathcal{F}_M$ for all $A \in A$, if $S \in \mathcal{F}_M$ then $X - S \in \mathcal{F}_M$, if $S, T \in \mathcal{F}_M$ then $S \cap T \in \mathcal{F}_M$, and if $S \in \mathcal{F}_M$ then $S^\circ \in \mathcal{F}_M$ i.e., $\mathcal{F}_M$ is the least set containing $\{i(A) | A \in A\}$ and closed under complements, intersections and interiors. Let $\mathcal{F}_M^\circ$ be the set $\{S^\circ | S \in \mathcal{F}_M\}$. We have $\mathcal{F}_M^\circ = \mathcal{F}_M \cap T$. The following is the main theorem of this section.

Theorem 14 Partition Theorem. *Let $M = \langle X, T, i \rangle$ be a topological model. Then there exists a set $\{\mathcal{F}^\psi\}_{\psi \in \mathcal{L}}$ of finite stable splittings such that*

1. *$\mathcal{F}^\psi \subseteq \mathcal{F}_M^\circ$ and $X \in \mathcal{F}^\psi$, for all $\psi \in \mathcal{L}$,*
2. *if $U \in \mathcal{F}^\psi$ then $U^\psi = \{x \in U | x, U \models \psi\} \in \mathcal{F}_M$, and*
3. *if ϕ is a subformula of ψ then $\mathcal{F}^\phi \subseteq \mathcal{F}^\psi$ and $\mathcal{F}^\psi$ is a finite stable splitting for ϕ,*

where $\mathcal{F}_M$, $\mathcal{F}_M^\circ$ as above.

Proof. By induction on the structure of the formula ψ. In each step we take care to refine the partition of the induction hypothesis.

- If $\psi = A$ is an atomic formula, then $\mathcal{F}^A = \{X, \emptyset\} = \{i(\top), i(\bot)\}$, since T is stable for all atomic formulae. We also have $\mathcal{F}^A \subseteq \mathcal{F}_M^\circ$ and $X^A = i(A) \in \mathcal{F}_M$.
- If $\psi = \neg\phi$ then let $\mathcal{F}^\psi = \mathcal{F}^\phi$, since the statement of the proposition is symmetric with respect to negation. We also have that for an arbitrary $U \in \mathcal{F}^\psi$, $U^\psi = U^{\neg\phi}$.
- If $\psi = \chi \wedge \phi$, let
$$\mathcal{F}^\psi = \text{Cl}(\mathcal{F}^\chi \cup \mathcal{F}^\phi).$$
Observe that $\mathcal{F}^\chi \cup \mathcal{F}^\phi \subseteq \mathcal{F}^{\chi \wedge \phi}$.
Now, if $W_i \in \mathcal{F}^\psi$, then there exists $U_j \in \mathcal{F}^\chi$ and $V_k \in \mathcal{F}^\phi$ such that

$$W_i = U_j \cap V_k \quad \text{and} \quad \text{Rem}^{\mathcal{F}^\psi} W_i \subseteq \text{Rem}^{\mathcal{F}^\chi} U_j \cap \text{Rem}^{\mathcal{F}^\phi} V_k$$

(e.g., $U_j = \bigcap\{U_m | W_i \subseteq U_m, \ U_m \in \mathcal{F}^\chi\}$ and $V_k = \bigcap\{V_n | W_i \subseteq V_n, \ V_n \in \mathcal{F}^\phi\}$.) Since $\text{Rem}^{\mathcal{F}^\chi} U_j$ is stable for χ and $\text{Rem}^{\mathcal{F}^\phi} V_n$ is stable for ϕ, their intersection is stable for $\chi \wedge \phi = \psi$, by Proposition 11(b), and so is its subset $\text{Rem}^{\mathcal{F}^\psi} W_i$, by Proposition 11(a). Thus $\mathcal{F}^\psi$ is a finite stable splitting for ψ containing X.
We have that $\mathcal{F}^\psi \subseteq \mathcal{F}_M$ whenever $\mathcal{F}^\chi \subseteq \mathcal{F}_M$ and $\mathcal{F}^\phi \subseteq \mathcal{F}_M^\circ$. Finally, $W_i^\psi = U_j^\chi \cap V_k^\phi$.
- Suppose $\psi = K\phi$. Then, by induction hypothesis, there exists a finite stable splitting $\mathcal{F}^\phi = \{U_1, \ldots, U_n\}$ for ϕ containing X. Let

$$W_i = (U_i^\phi)^\circ,$$

for all $i \in \{1, \ldots, n\}$.

Observe that if $x \in U_i - W_i$ then $x, V \models \neg\phi$, for all $V \in \text{Rem}^{\mathcal{F}^{\phi}} U_i$ and $x \in V$, since $\text{Rem}^{\mathcal{F}^{\phi}} U_i$ is stable for ϕ, by induction hypothesis.

Now, if $V \in \text{Rem}^{\mathcal{F}^{\phi}} U_i \cap \downarrow W_i$, for some $i \in \{1 \ldots, n\}$, then $x, V \models \phi$ for all $x \in V$, by definition of W_i, hence $x, V \models K\phi$ for all $x \in V$.

On the other hand, if $V \in \text{Rem}^{\mathcal{F}^{\phi}} U_i - \downarrow W_i$ then there exists $x \in V$ such that $x, V \models \neg\phi$ (otherwise $V \subseteq W_i$). Thus we have $x, V \models \neg K\phi$ for all $x \in V$. Hence $\text{Rem}^{\mathcal{F}^{\phi}} U_i \cap \downarrow W_i$ and $\text{Rem}^{\mathcal{F}^{\phi}} U_i - \downarrow W_i$ are stable for $K\phi$. Thus, the set

$$F = \{\text{Rem}^{\mathcal{F}} U_i \mid W_i \notin \text{Rem}^{\mathcal{F}} U_i\} \cup \{\text{Rem}^{\mathcal{F}} U_j - \downarrow W_j, \text{Rem}^{\mathcal{F}} U_j \cap \downarrow W_j \mid W_j \in U_j\}$$

is a partition of T and its members are stable for $K\phi$. Let $\sim_F$ be the equivalence relation on T induced by F and let

$$\mathcal{F}^{K\phi} = \text{Cl}(\mathcal{F}^{\phi} \cup \{W_i \mid W_i \in \text{Rem}^{\mathcal{F}^{\phi}} U_i\}).$$

We have that $\mathcal{F}^{K\phi}$ is a finite set of opens and $\mathcal{F}^{\phi} \subseteq \mathcal{F}^{K\phi}$. Thus, $\mathcal{F}^{K\phi}$ is finite and contains X. We have only to prove that $\mathcal{F}^{\overline{K}\phi}$ is a stable splitting for $K\phi$, i.e., every remainder of an open in $\mathcal{F}^{K\phi}$ is stable for $K\phi$.

If $V_1 \not\sim_F V_2$, where $V_1, V_2 \in T$, then there exists $U = U_i$ or W_i for some $i = 1, \ldots, n$ such that $V_1 \subseteq U$ while $V_2 \not\subseteq U$. But this implies that $V_1 \not\sim_{\mathcal{F}^{K\phi}} V_2$. Therefore $\{\text{Rem}^{\mathcal{F}^{K\phi}} U\}_{U \in \mathcal{F}^{K\phi}}$ is a refinement of F and $\mathcal{F}^{K\phi}$ is a finite stable splitting for $K\phi$ using Proposition 11(a).

We have that $\mathcal{F}^{K\phi} \subseteq \mathcal{F}^{\circ}_{\mathcal{M}}$ because $W_i \in \mathcal{F}^{\circ}_{\mathcal{M}}$, for $i = 1, \ldots, n$. Now if $U \in \mathcal{F}^{\psi}$ then either $U^{K\phi} = U$ or $U^{K\phi} = \emptyset$.

- Suppose $\psi = \Box\phi$. Then, by induction hypothesis, there exists a finite stable splitting $\mathcal{F}^{\phi} = \{U_1, \ldots, U_n\}$ for ϕ containing X. Let

$$\mathcal{F}^{\Box\phi} = \text{Cl}(\mathcal{F}^{\phi} \cup \{U_i \Rightarrow U_j \mid 1 \leq i, j \leq n\}),$$

where $\Rightarrow$ is the implication of the complete Heyting algebra T i.e., $V \subseteq U \Rightarrow W$ if and only if $V \cap U \subseteq W$ for $V, U, W \in T$. We have that $U \Rightarrow W$ equals $(X - (U - W))^{\circ}$. Clearly, $\mathcal{F}^{\Box\phi}$ is a finite splitting containing X and $\mathcal{F}^{\phi} \subseteq \mathcal{F}^{\Box\phi}$. We have only to prove that $\mathcal{F}^{\Box\phi}$ is stable for $\Box\phi$. But first, we prove the following claim:

Claim 15. Suppose $U \in \mathcal{F}^{\phi}$ and $U' \in \mathcal{F}^{\Box\phi}$. Then

$$U' \cap U \in \text{Rem}^{\mathcal{F}^{\phi}} U \iff V \cap U \in \text{Rem}^{\mathcal{F}^{\phi}} U \text{ for all } V \in \text{Rem}^{\mathcal{F}^{\Box\phi}} U'.$$

Proof. The one direction is straightforward. For the other, let $V \in \text{Rem}^{\mathcal{F}^{\Box\phi}} U'$ and suppose $V \cap U \notin \text{Rem}^{\mathcal{F}^{\phi}} U$ towards a contradiction. This implies that there exists $U'' \in \mathcal{F}^{\phi}$, with $U'' \subset U$, such that $V \cap U \subseteq U''$. Thus, $V \subseteq U \Rightarrow U''$ but $U' \not\subseteq U \Rightarrow U''$. But $U \Rightarrow U'' \in \mathcal{F}^{\Box\phi}$ which contradicts $U' \sim_{\mathcal{F}^{\Box\phi}} V$, by Proposition 8.

Let $U' \in \mathcal{F}^{\Box\phi}$. We must prove that $\mathrm{Rem}^{\mathcal{F}^{\Box\phi}} U'$ is stable for $\Box\phi$.
Suppose that $x, U' \models \neg\Box\phi$. We must prove that

$$x, V' \models \neg\Box\phi$$

for all $V' \in \mathrm{Rem}^{\mathcal{F}^{\Box\phi}} U'$ such that $x \in V'$.
Since $x, U' \models \neg\Box\phi$, there exists $V \in \mathcal{T}$, with $x \in V$ and $V \subseteq U'$, such
that $x, V \models \neg\phi$. Since $\mathcal{F}^\phi$ is a splitting, there exists $U \in \mathcal{F}^\phi$ such that
$V \in \mathrm{Rem}^{\mathcal{F}^\phi} U$. Observe that $V \subseteq U' \cap U \subseteq U$, so $U' \cap U \in \mathrm{Rem}^{\mathcal{F}^\phi} U$,
by Proposition 5(c).
By Claim 15, for all $V' \in \mathrm{Rem}^{\mathcal{F}^{\Box\phi}} U'$, we have $V' \cap U \in \mathrm{Rem}^{\mathcal{F}^\phi} U$. Thus if
$x \in V'$ then $x, V' \cap U \models \neg\phi$, because $\mathrm{Rem}^{\mathcal{F}^\phi} U$ is stable for ϕ, by induction
hypothesis. This implies that, for all V' such that $V' \in \mathrm{Rem}^{\mathcal{F}^{\Box\phi}} U'$ and $x \in V$,
we have $x, V' \models \neg\Box\phi$.
Therefore, $\mathcal{F}^{\Box\phi}$ is a finite stable splitting for $\Box\phi$.
Now $U_i \Rightarrow U_j \in \mathcal{F}^\circ_{\mathcal{M}}$ for $1 \le i, j \le n$, hence $\mathcal{F}^{\Box\phi} \subseteq \mathcal{F}^\circ_{\mathcal{M}}$.
Finally, let U belong to $\mathcal{F}^{\Box\phi}$ and $V_1, \ldots, V_m$ be all opens in $\mathcal{F}^\phi$ such that
$U \cap V_i \in \mathrm{Rem}^{\mathcal{F}^\phi} V_i$, for $i = 1, \ldots, m$. Then $x, U \models \Diamond\neg\phi$ if and only if there
exists $j \in \{1, \ldots, m\}$ with $x \in V_j$ and $x, V_j \models \neg\phi$ because $x, V_j \cap U \models \neg\phi$ since
$V_j \cap U \in \mathrm{Rem}^{\mathcal{F}^\phi} V_j$. This implies that

$$U^{\neg\Box\phi} = U^{\Diamond\neg\phi} = U \cap \bigcup_{i=1}^m V_i^{\neg\phi}.$$

Since $U, V_1^{\neg\phi}, \ldots, V^{\neg\phi}$ belong to $\mathcal{F}_{\mathcal{M}}$, so does $U^{\neg\Box\phi}$ and, therefore, $U^{\Box\phi} = U - U^{\neg\Box\phi}$.

In all steps of induction we refine the finite splitting, so if ϕ is a subformula
of ψ, then $\mathcal{F}^\phi \subseteq \mathcal{F}^\psi$ and $\mathcal{F}^\psi$ is stable for ϕ using Proposition 11(a).

Theorem 14 gives us a great deal of intuition for topological models. It describes in detail the expressible part of the topological lattice for the completeness result as it appears in (Georgatos 1993) and paves the road for the reduction
of the theory of topological models to that of spatial lattices and the decidability
result of this section.

4 Basis Model

Let $\mathcal{T}$ be a topology on a set X and $\mathcal{B}$ a basis for $\mathcal{T}$. We denote satisfaction in
the models $\langle X, \mathcal{T}, i \rangle$ and $\langle X, \mathcal{B}, i \rangle$ by $\models_{\mathcal{T}}$ and $\models_{\mathcal{B}}$, respectively. In the following
proposition we prove that each equivalence class under $\sim_{\mathcal{F}}$ contains an element
of a basis closed under finite unions.

Proposition 16. *Let $(X, \mathcal{T})$ be a topological space, and let $\mathcal{B}$ be a basis for $\mathcal{T}$
closed under finite unions. Let $\mathcal{F}$ be any finite subset of $\mathcal{T}$. Then for all $V \in \mathcal{F}$
and all $x \in V$, there is some $U \in \mathcal{B}$ with $x \in U \subseteq V$ and $U \in \mathrm{Rem}^{\mathcal{F}} V$.*

Proof. By finiteness of $\mathcal{F}$, let $V_1, \ldots, V_k$ be the elements of $\mathcal{F}$ such that $V \not\subseteq V_i$, for $i \in \{1, \ldots, k\}$. Since $V_i \neq V$, take $x_i \in V - V_i$ for $i \in \{1, \ldots, k\}$. Since $\mathcal{B}$ is a basis for $\mathcal{T}$, there exist U_x, U_i, with $x \in U_x$ and $x_i \in U_i$, such that U_x and U_i are subsets of V for $i \in \{1, \ldots, k\}$. Set

$$U = (\bigcup_{i=1}^{k} U_i) \cup U_x.$$

Observe that $x \in U$, and $U \in \mathcal{B}$, as it is a finite union of members of $\mathcal{B}$. Also $U \in \mathrm{Rem}^{\mathcal{F}} V$, since $U \in \downarrow V$ but $U \notin \bigcup \downarrow V_i$ for $i \in \{1, \ldots, k\}$.

Corollary 17. *Let $(X, \mathcal{T})$ be a topological space, $\mathcal{B}$ a basis for $\mathcal{T}$ closed under finite unions, $x \in X$ and $U \in \mathcal{B}$. Then*

$$x, U \models_{\mathcal{T}} \phi \iff x, U \models_{\mathcal{B}} \phi.$$

Proof. By induction on ϕ.

The interesting case is when $\phi = \Box \psi$. Fix x, U, and ψ. By Proposition 14, there exists a finite stable splitting $\mathcal{F}$ for ϕ and its subformulae such that $\mathcal{F}$ contains X and U. Assume that $x, U \models_{\mathcal{B}} \Box \psi$, and $V \in \mathcal{T}$ such that $V \subseteq U$. By Proposition 5(b), there is some $V' \subseteq U$ in $\mathcal{F}$ with $V \in \mathrm{Rem}^{\mathcal{F}} V'$. By Proposition 16, let $W \in \mathcal{B}$ be such that $W \in \mathrm{Rem}^{\mathcal{F}} V'$ with $x \in W$. So $x, W \models_{\mathcal{B}} \psi$, and thus by induction hypothesis, $x, W \models_{\mathcal{T}} \psi$. By stability, twice, $x, V \models_{\mathcal{T}} \psi$ as well.

We are now going to prove that a model based on a topological space $\mathcal{T}$ is equivalent to the one induced by any basis of $\mathcal{T}$ which is lattice. Observe that this enables us to reduce the theory of topological spaces to that of spatial lattices and, therefore, to answer the conjecture of (Moss and Parikh 1992): a completeness theorem for subset spaces which are lattices will extend to the smaller class of topological spaces.

Theorem 18. *Let $(X, \mathcal{T})$ be a topological space and $\mathcal{B}$ a basis for $\mathcal{T}$ closed under finite unions. Let $\mathcal{M}_1 = \langle X, \mathcal{T}, i \rangle$ and $\mathcal{M}_2 = \langle X, \mathcal{B}, i \rangle$ be the corresponding models. Then, for all ϕ,*

$$\mathcal{M}_1 \models \phi \iff \mathcal{M}_2 \models \phi.$$

Proof. It suffices to prove that $x, U \models_{\mathcal{T}} \phi$, for some $U \in \mathcal{T}$, if and only if $x, U' \models_{\mathcal{B}} \phi$, for some $U' \in \mathcal{B}$.

Suppose $x, U \models_{\mathcal{T}} \phi$, where $U \in \mathcal{T}$, then, by Corollary 17, there exists $U' \in \mathcal{B}$ such that $x \in U'$ and $x, U \models_{\mathcal{T}} \phi$. By Corollary 17, $x, U' \models_{\mathcal{B}} \phi$.

Suppose $x, U \models_{\mathcal{B}} \phi$, where $U \in \mathcal{B}$, then $x, U \models_{\mathcal{T}} \phi$, by Corollary 17.

5 Finite Satisfiability

Proposition 19. *Let $\langle X, T \rangle$ be a subset space. Let $\mathcal{F}$ be a finite stable splitting for a formula ϕ and all its subformulae, and assume that $X \in \mathcal{F}$. Then for all $U \in \mathcal{F}$, all $x \in U$, and all subformulae ψ of ϕ, $x, U \models_T \psi$ iff $x, U \models_{\mathcal{F}} \psi$.*

Proof. The argument is by induction on ϕ. The only interesting case to consider is when $\phi = \square \psi$.

Suppose first that $x, U \models_{\mathcal{F}} \square \psi$ with $U \in \mathcal{F}$. We must show that $x, U \models_T \square \psi$ also. Let $V \in T$ such that $V \subseteq U$; we must show that $x, V \models_T \psi$. By Proposition 5(b), there is some $V' \subseteq U$ in $\mathcal{F}$ with $V \in \mathsf{Rem}^{\mathcal{F}} V'$. So $x, V' \models_{\mathcal{F}} \psi$, and by induction hypothesis, $x, V' \models_T \psi$. By stability, $x, V \models_T \psi$ also.

The other direction (if $x, U \models_T \square \psi$, then $x, U \models_{\mathcal{F}} \square \psi$), is an easy application of the induction hypothesis.

Constructing the quotient of T under $\sim_{\mathcal{F}}$ is not adequate for generating a finite model because there may still be an infinite number of points. It turns out that we only need a finite number of them.

Let $\mathcal{M} = \langle X, T, i \rangle$ be a topological model, and define an equivalence relation $\sim$ on X by $x \sim y$ iff

(a) for all $U \in T$, $x \in U$ iff $y \in U$, and
(b) for all atomic A, $x \in i(A)$ iff $y \in i(A)$.

Further, denote by x^* the equivalence class of x, and let $X^* = \{x^* : x \in X\}$. For every $U \in T$ let $U^* = \{x^* : x \in X\}$, then $T^* = \{U^* : U \in T\}$ is a topology on X^*. Define a map i^* from the atomic formulae to the powerset of X^* by $i^*(A) = \{x^* : x \in i(A)\}$. The entire model $\mathcal{M}$ lifts to the model $\mathcal{M}^* = \langle X^*, T^*, i^* \rangle$ in a well-defined way.

Lemma 20. *For all x, U, and ϕ,*

$$x, U \models_{\mathcal{M}} \phi \qquad \text{iff} \qquad x^*, U^* \models_{\mathcal{M}^*} \phi .$$

Proof. By induction on ϕ.

Theorem 21. *If ϕ is satisfied in any topological space, then ϕ is satisfied in a finite topological space.*

Proof. Let $\mathcal{M} = \langle X, T, i \rangle$ be such that for some $x \in U \in T$, $x, U \models_{\mathcal{M}} \phi$. Let $\mathcal{F}^{\phi}$ be a finite stable splitting (by Theorem 14) for ϕ and its subformulae with respect to $\mathcal{M}$. By Proposition 19, $x, U \models_{\mathcal{N}} \phi$, where $\mathcal{N} = \langle X, \mathcal{F}, i \rangle$. We may assume that $\mathcal{F}$ is a topology, and we may also assume that the overall language has only the (finitely many) atomic symbols which occur in ϕ. Then the relation $\sim$ has only finitely many classes. So the model $\mathcal{N}^*$ is finite. Finally, by Lemma 20, $x^*, U^* \models_{\mathcal{N}^*} \phi$.

158

Observe that the finite topological space is a quotient of the initial one under two equivalences. The one equivalence is $\sim_{\mathcal{F}}^{\phi}$ on the open subsets of the topological space, where $\mathcal{F}^{\phi}$ is the finite splitting corresponding to ϕ and its cardinality is a function of the complexity of ϕ. The other equivalence is $\sim_X$ on the points of the topological space and its number of equivalence classes is a function of the atomic formulae appearing in ϕ. The following simple example shows how a topology is formed with the quotient under these two equivalences

Example 2. Let X be the interval $[0,1)$ of the real line with the set

$$\mathcal{T} = \{\emptyset\} \cup \{ [0, \frac{1}{2^n}) \mid n = 0,1,2,\dots \}$$

as topology. Suppose that we have only one atomic formula, call it A, such that $i(A) = \{0\}$. Then it is easy to see that the model $\langle X, \mathcal{T}, i \rangle$ is equivalent to the finite topological model $\langle X^*, \mathcal{T}^*, i^* \rangle$, where

$$X^* = \{ x_1, x_2 \},$$
$$\mathcal{T}^* = \{ \emptyset, \{x_1, x_2\} \}, \text{ and}$$
$$i(A) = \{ x_1 \}.$$

So the overall size of the (finite) topological space is bounded by a function of the complexity of ϕ. Thus if we want to test whether a given formula is invalid we have a finite number of finite topological spaces where we have to test its validity. Thus we have the following

Theorem 22. *The theory of topological spaces is decidable.*

Observe that the last two results apply for lattices of subsets by Theorem 18.

Acknowledgements

I wish to thank Larry Moss and Rohit Parikh for helpful comments and suggestions.

References

Fagin, R., J.Y. Halpern, and M.Y. Vardi: "A model-theoretic analysis of knowledge," in: *Journal of the Association for Computing Machinery*, 38 (2) (1991) 382–428.

Georgatos, K.: *Modal Logics for Topological Spaces.* Ph.D. Dissertation. City University of New York, 1993.

Gierz, G., K.H. Hoffman, K. Keimel, J.D. Lawson, M.W. Mislove, and D.S. Scott: *A Compendium of Continuous Lattices.* Berlin, Heidelberg: Springer-Verlag, 1980.

Halpern, J.Y., and Y. Moses: "Knowledge and common knowledge in a distributed environment," in: *Proceedings of the Third ACM Symposium on Principles of Distributed Computing* (1984) 50-61.

Hintikka, J.: *Knowledge and Belief.* Ithaca, New York: Cornell University Press, 1962.

Moss. L.S., and R. Parikh: "Topological reasoning and the logic of knowledge," in: Y. Moses (ed.), *Proceedings of the Fourth Conference (TARK 1992)* (1992) 95-105.

Parikh, R., and R. Ramanujam: "Distributed computing and the logic of knowledge," in: R. Parikh (ed.) *Logics of Programs*, number 193 in Lecture Notes in Computer Science, Berlin, New York: Springer Verlag (1985) 256-268.

Rasiowa, H., and R. Sikorski: *The Mathematics of Metamathematics.* Panstwowe Wydawnictwo Naukowe, Warszawa, Poland, second edition, 1968.

Vickers, S.: *Topology via Logic.* Cambridge Studies in Advanced Computer Science. Cambridge: Cambridge University Press, 1989.

Rough Logic for Multi-agent Systems

Cecylia M. Rauszer

Institute of Mathematics, Warsaw University
ul. Banacha 2, 02-097 Warsaw, Poland
E-mail: rauszer@mimuw.edu.pl

Abstract. Reasoning about knowledge is one of the most challenging problems in philosophy, artificial intelligence and logic. Various approaches have been presented, such as nonmonotonic reasoning, fuzzy sets, probabilistic logic, and so on. In this paper we focus on *rough set philosophy* to describe reasoning about knowledge, which is understood as the ability to classify objects. Rough sets are used as a mathematical tool to deal with uncertain and imprecise data. The aim of this paper is to present a complete formal system for reasoning based on incomplete information of a multi-agent system. The logic under consideration is a multi modal logic that can be used as a query language for groups of agents.

1 Introduction

The concept of rough sets has been proposed (Pawlak 1982) as a new mathematical tool to deal with uncertain and imprecise data. In the paper we focus on the rough sets philosophy to model reasoning about knowledge (cf. (Pawlak 1982), (Orłowska 1989), (Rasiowa 1990), (Rauszer 1992a, to appear)).

In our approach, knowledge is understood as the ability to classify objects which are taken from the fixed set U (called the universe of discourse, or shortly *universe*). Any classification is in fact a partition of the universe. Objects being in the same class are indiscernible by means of knowledge provided by the classification. For instance, if $U = \{o_1, o_2, o_3, o_4, o_5\}$ then the classifications (partitions) $\mathcal{E}_s = \{\{o_1, o_2, o_3\}, \{o_4\}, \{o_5\}\}$ and $\mathcal{E}_t = \{\{o_1, o_2\},$ $\{o_3, o_4\}, \{o_5\}\}$ are examples of knowledge about the universe U. More precisely, $\mathcal{E}_s$ and $\mathcal{E}_t$ given above are treated as knowledge base of agents s and t about the universe U. In the example, the agent s identifies objects o_1, o_2, o_3 and distinguishes objects o_4 and o_5, whereas the agent t identifies o_1 with o_2 and o_3 with o_4 and recognizes o_5.

We will discuss the multi-agent case. We assume that a family of groups of agents T is given and each group perceives the same universe of discourse U. Sometimes, for simplicity, we will refer to elements of T as agents, not groups of agents.

Let $\mathcal{E} = \{\mathcal{E}_t\}_{t \in T}$ be a family of partitions of U, that is, each $\mathcal{E}_t$ is considered as knowledge base of an agent t (of a group of agents t) about U. We extend our semantics so that we can reason about distributed knowledge which is intended to convey information about knowledge of every agent from the group of agents.

We assume that each group of agents can have two kinds of distributed knowledge base, strong and weak.

We say that a group of agents has a *strong distributed knowledge base*, provided objects which are distinguishable for one agent from the group are also distinguishable for the whole group, that is, an agent t with better knowledge plays a dominating role in the group. More precisely, if $\mathcal{E}_s$ is interpreted as knowledge base of an agent s and $\mathcal{E}_t$ as knowledge base of an agent t about the same universe U then $\mathcal{E}_{s \vee t}$ may be viewed as a strong distributed knowledge base of the group $\{s, t\}$, denoted by st, which intuitively may be described as follows: If an agent s knows that an object x is different from objects $x_{j_1}, \ldots, x_{j_m}$, then x has to be different from those objects in $\mathcal{E}_{s \vee t}$. As a consequence we have: if an object x is distinguished from y in $\mathcal{E}_{s \vee t}$, that is, the block $[x]$ is different from the block $[y]$ (in $\mathcal{E}_{s \vee t}$) then it means that at least one of the agents s, t distinguishes x from y. If objects x and y are indiscernible in $\mathcal{E}_{s \vee t}$, then it means that each agent from the group st, according to her knowledge, is not able to distinguish between x and y. For instance, in the above example the partition $\mathcal{E}_s$ better classifies an object o_4 than knowledge $\mathcal{E}_t$. Hence, the class $\{o_4\}$ has to appear in $\mathcal{E}_{s \vee t}$ despite of, using knowledge base $\mathcal{E}_t$ objects o_3 and o_4 are indiscernible.

The notion of a *weak distributed knowledge base* may be interpreted as a classification that preserves less efficient knowledge. This means that if an agent, say t, identifies some objects, then they are indiscernible in each group to which the agent t belongs. If $\mathcal{E}_s$ and $\mathcal{E}_t$ denote knowledge bases of agents s and t, respectively, then we denote by $\mathcal{E}_{s \wedge t}$ a weak distributed knowledge base of the group st. Thus, if $[x] = \{x\} \in \mathcal{E}_{s \wedge t}$, then we infer that there is a consensus of agents s and t about the object x, i.e., $[x] \in \mathcal{E}_s \cap \mathcal{E}_t$. If $\{x_1, \ldots, x_n\} \in \mathcal{E}_{s \wedge t}$, then it means that either there is consensus of the agents s and t about $x_1, \ldots, x_n$ or for any subset of $\{x_1, \ldots, x_n\}$ there is no consensus. For instance, in the above example because of the objects o_3 and o_4 are indiscernible by the agent t, they have to remain indiscernible in $\mathcal{E}_{s \wedge t}$. Moreover, because of o_3 is not distinguished from objects o_1, o_2 by an agent s, the block $\{o_1, o_2, o_3, o_4\}$ is an element of the partition $\mathcal{E}_{s \wedge t}$.

It is shown (in Section 2) that if the family $\mathcal{E} = \{\mathcal{E}_t\}_{t \in T}$ is closed with respect to strong and weak distributed knowledge, then it is a lattice.

In Section 3 we examine lower and upper approximations of sets of objects (Pawlak 1992). Intuitively speaking, by a lower approximation of X determined by a partition of U we mean the set of objects of U that without any doubt belong to X. An upper approximation of X is a set of objects which could be classified as elements of X. Finally, we consider the boundary of X which is in a sense undecidable area of the universe.

Let $\mathcal{E}_t$ be a knowledge base of an agent t and let R_t denote an equivalence relation determined by $\mathcal{E}_t$. Then the t-lower approximation of a set of objects $X \subseteq U$, denoted by $\underline{R_t}(X)$, is intended as knowledge of an agent t about X. In other words, $\underline{R_t}(X)$ is read: *an agent t knows X*. Hence, $\underline{R_t}(X)$ consists of objects which are classified by an agent t as those which, according to knowledge base $\mathcal{E}_t$, have a property given by X. If $\underline{R_t}(X) = X$, then we say that X is *t-defined*.

If $\underline{R}_t(X) \neq X$ then X is *t-rough*. In the former case, the agent t can justify if any object from the universe U belongs to X or to $-X$. In the latter case, the t-boundary of X is a nonempty set, that means that the agent t is unable to decide the membership problem for X.

The next problem is to find methods which allow to represent knowledge in terms of classifications. It turns out that information systems introduced by Pawlak (1982, 1991) provide good models of knowledge. The idea is that information systems may be used as "real" description of a knowledge base. In Section 4 we recall some basic properties of information systems, describing them as tables with columns labelled by attributes and rows labelled by objects. Each row in the data table represents information about a corresponding object. In general, in a given information system we are not able to distinguish all single objects (using attributes provided by the system), since objects can have the same values on some attributes. As a consequence, any set of attributes establishes a partition of the set U of all objects. Suppose now, that S_t is an information system describing the universe U by means of a set of attributes $\mathbf{A}_t$. Denote it by $(U, \mathbf{A}_t)$, that is, $S_t = (U, \mathbf{A}_t)$ and let $\mathcal{E}_{\mathbf{A}_t}$ be the partition given by $\mathbf{A}_t$. If $\mathcal{E}_{\mathbf{A}_t} = \mathcal{E}_t$ then we say that S_t represents knowledge base $\mathcal{E}_t$ of an agent t and call S_t a *knowledge representation system* of $\mathcal{E}_t$. Clearly, such a representation is not unique.

We show that if $S_s = (U, \mathbf{A}_s)$ is a knowledge representation system of $\mathcal{E}_s$ and $S_t = (U, \mathbf{A}_t)$ is a knowledge representation system of $\mathcal{E}_t$, then the system $(U, \mathbf{A}_s \cup \mathbf{A}_t)$ represents a strong distributed knowledge base of the group st i.e., $\mathcal{E}_{s \vee t} = \mathcal{E}_{\mathbf{A}_s \cup \mathbf{A}_t}$. In the case of a weak distributed knowledge base, in general we have $\mathcal{E}_{s \wedge t} \neq \mathcal{E}_{\mathbf{A}_s \cap \mathbf{A}_t}$. Clearly, if $\mathcal{E}_{s \wedge t}$ is a weak distributed knowledge base of $\mathcal{E}_s$ and $\mathcal{E}_t$, then there is a knowledge representation system $(U, \mathbf{A})$ such that the partition determined by $\mathbf{A}$ is equal to $\mathcal{E}_{s \wedge t}$. However, the set of attributes $\mathbf{A}$ need not to be the same as the set $\mathbf{A}_s \cap \mathbf{A}_t$.

It is also shown in Section 4 that the family $\{S\}_{t \in T}$ of knowledge representation systems may be treated as a lattice. The intuitive meaning of the lattice ordering $\leq$ on this family is that, if $S_s \leq S_t$, then the sharpness of perception of U and therefore feature recognitions of objects from the universe U by an agent s is weaker than that of an agent t.

In Section 5 we introduce and examine a formal system called *rough logic*. It is intended as a logic which reflects properties of approximations of knowledge plus some other features of knowledge representation systems. It seems that the logic we have been considering may provide good approximations to reasoning carried out by a knowledge base (Rauszer 1992a). Roughly speaking, our logic is a modal logic with a finite number of modal operators I_t. Each modal operator corresponds to knowledge represented by an information system. A formula of the form $I_t \phi$ might be read as *an agent t knows ϕ*. In other words, $I_t \phi$ represents the t-lower approximation of the set of objects which satisfy ϕ in an information system $S_t = (U, \mathbf{A}_t)$, that is, $I_t \phi$ represents the set of all positive instances of ϕ.

In Section 6 we present a theorem about knowledge understood as a partition.

Finally, we show that a formula of the form $I_{s \vee t} \phi$ may be treated as distributed knowledge in the sense of (Halpern and Moses 1992). Moreover, if knowledge of an agent t is understood as the set of all positive and negative instances of ϕ, that is, as a formula of the form $I_t\phi \vee I_t - \phi$, then knowledge of an agent is not closed under modus ponens and also some paradoxes of epistemic logic are eliminated. In particular, *what an agent knows is true* and *an agent always knows all the consequences of her knowledge* are not theorems of the logic under consideration.

2 Knowledge Base

In our approach, knowledge is understood as the ability to classify objects. Objects are treated as elements of a real or abstract world called the *universe of discourse* (or *universe*). In our understanding, knowledge concerns the classification of parts of the universe.

We explain this idea more precisely.

Let T be a family of groups o agents that perceive the same universe of discourse U. Any subset X of U is said to be a *concept*. For every $t \in T$, let $\mathcal{E}_t$ denote a partition of U established by a group of agents t. The family $\mathcal{E}_t$ is said to be a *knowledge base of a group t of agents about the universe U*, or simply, a *knowledge base of an agent t*.

Let T be a family of agents (groups of agents) and let for every $t \in T$, $\mathcal{E}_t$ be a knowledge base of an agent t about U. For simplicity, each block from $\mathcal{E}_t$ is denoted by $[x]_t$, instead of $[x]_{\mathcal{E}_t}$. Each block $[x]_t$ is a concept and we call it a *basic concept*. If a basic concept $[x]_t \in \mathcal{E}_t$ contains more than one object, then objects from $[x]_t$ are not distinguishable with respect to the knowledge base $\mathcal{E}_t$.

Let as before, T denote a family of agents and let $\mathcal{E} = \{\mathcal{E}_t\}_{t \in T}$ be a family of all partitions determined by agents from T. Let $\prec$ be a binary relation on $\mathcal{E}$ defined as follows:

$$\mathcal{E}_s \prec \mathcal{E}_t \text{ if and only if } \forall[x]_s \exists[y]_t \ [x]_s \subseteq [y]_t \ .$$

It is not difficult to prove that $\prec$ is an ordering relation on $\mathcal{E}$, that is, $(\mathcal{E}, \prec)$ is a poset.

Denote by $\mathcal{E}_s \wedge \mathcal{E}_t$ the following set:

$$\{[x]_s \cap [y]_t : [x]_s \cap [y]_t \neq \emptyset\}.$$

It is easily proved that $\mathcal{E}_s \wedge \mathcal{E}_t$ is a partition of U and we call this partition a *strong distributed knowledge base* of a group of agents ts. Moreover, if $\mathcal{E}_s \wedge \mathcal{E}_t$ belongs to $\mathcal{E}$ then it is the *infimum (inf)* of $\mathcal{E}_s$ and $\mathcal{E}_t$ in $(\mathcal{E}, \prec)$.

Observe that if $\mathcal{E}$ is a family of all partitions of the universe U, then $\mathcal{E}$ is closed with respect to $\wedge$.

Notice also:

Lemma 2.1 *If $\mathcal{E}$ is closed with respect to $\wedge$ then $(\mathcal{E}, \wedge)$ is a lower semi-lattice.*

□

Now put

$$\mathcal{E}_s \vee \mathcal{E}_t = \{X \subseteq U : X \text{ is the union of } \mathbf{all} \ [x]_s \text{ and } [y]_t \text{ such that } [x]_s \cap [y]_t \neq \emptyset\}.$$

It is not difficult to show that $\mathcal{E}_s \vee \mathcal{E}_t$ is a partition of U. We call this partition a *weak distributed knowledge base* of a group of agents ts. Moreover, if the partition $\mathcal{E}_s \vee \mathcal{E}_t$ exists in $\mathcal{E}$, then $\mathcal{E}_s \vee \mathcal{E}_t$ is the *supremum (sup)* in $(\mathcal{E}, \prec)$ of $\mathcal{E}_s$ and $\mathcal{E}_t$.

Lemma 2.2 *If a family $\mathcal{E}$ is closed with respect to $\wedge$ and $\vee$, then the structure $(\mathcal{E}, \vee, \wedge)$ is a lattice with the zero element and the unit element.*

Proof: Clearly, the mentioned structure is a lattice. The strong distributed knowledge base of all $\mathcal{E}_s$'s, that is, the $inf\mathcal{E}$ is the zero element and the weak distributed knowledge base of all $\mathcal{E}_s$'s, that is, the $sup\mathcal{E}$ is the unit element in this lattice.

□

Notice, that if $\mathcal{E}$ is the family of all partitions of the universe U then $(\mathcal{E}, \vee, \wedge, \mathbf{0}, \mathbf{1})$ is a lattice. We call every sublattice of the lattice $(\mathcal{E}, \vee, \wedge, \mathbf{0}, \mathbf{1})$ a *lattice of partitions*. The example given below shows that a lattice of partitions needs not to be distributive.

Example 1. Let U consist of five objects $\{o_1, o_2, o_3, o_4, o_5\}$ and let $\mathcal{E}_s = \{\{o_1, o_2\}, \{o_3\}, \{o_4\}, \{o_5\}\}$, $\mathcal{E}_t = \{\{o_1, o_4\}, \{o_2, o_3\}, \{o_5\}\}$ and $\mathcal{E}_w = \{\{o_1, o_2\}, \{o_3, o_4\}, \{o_5\}\}$. Then a strong distributed knowledge base $\mathcal{E}_s \wedge \mathcal{E}_t$ is the following partition of U: $\{\{o_1\}, \{o_2\}, \{o_3\}, \{o_4\}, \{o_5\}\}$, and a weak distibuted knowledge base $\mathcal{E}_s \vee \mathcal{E}_t$ is the partition $\{\{o_1, o_2, o_3, o_4\}, \{o_5\}\}$. Notice that $\mathcal{E}_w \wedge (\mathcal{E}_s \vee \mathcal{E}_t) = \mathcal{E}_w$ and $\mathcal{E}_w \wedge \mathcal{E}_s \vee \mathcal{E}_w \wedge \mathcal{E}_t = \mathcal{E}_s$. Hence, the family $\{\mathcal{E}_s, \mathcal{E}_t, \mathcal{E}_w, \mathcal{E}_s \vee \mathcal{E}_t, \mathcal{E}_s \wedge \mathcal{E}_t\}$ is an example of a non-distributive lattice of partitions. $\mathcal{E}_s \wedge \mathcal{E}_t$ is the zero element and $\mathcal{E}_t \vee \mathcal{E}_t$ is the unit element in this lattice.

□

We finish this section with the following remark. Let T be a family of groups of agents such that $\mathcal{E} = \{\mathcal{E}_t\}_{t \in T}$ is a lattice of partitions, where each $\mathcal{E}_t$ is referred to us as a knowledge base of an agent $t \in T$ about the universe U. It might be easily shown that the relation $\leq$ defined on T as follows:

$$s \leq t \text{ if and only if } \mathcal{E}_t \prec \mathcal{E}_t,$$

is an ordering relation on T. Thus $(T, \leq)$ is a poset. Moreover, one can also prove that $sup\{s, t\}$, denoted by $s \vee t$, and $inf\{s, t\}$, denoted by $s \wedge t$, exist in $(T, \leq)$. Namely, we have

$$s \vee t = sup\{s, t\} \text{ if and only if } \mathcal{E}_{s \vee t} = \mathcal{E}_s \wedge \mathcal{E}_t$$

and

$$s \wedge t = inf\{s, t\} \text{ if and only if } \mathcal{E}_{s \wedge t} = \mathcal{E}_s \vee \mathcal{E}_t.$$

Thus we conclude:

Lemma 2.3 $(T, \vee, \wedge)$ *is a lattice with the zero and unit element provided* $\mathcal{E} = \{\mathcal{E}_t\}_{t \in T}$ *is a lattice of partitions.*

□

3 Lower and Upper Approximation

Observe that some concepts may be expressed as the set-theoretical union of certain basic concepts in a knowledge base about U, but they cannot be defined as the union of basic concepts from another knowledge base. Hence, if a concept cannot be covered by basic concepts from a given knowledge base $\mathcal{E}_t$, then the question arises whether it can be "approximately" defined by $\mathcal{E}_t$. In this section we are going to discuss this problem.

Let $\mathcal{E}_t$ be a knowledge base of an agent t about U and let R_t be the equivalence relation determined by $\mathcal{E}_t$. With every concept X, $X \subseteq U$, we associate three sets: $\underline{R}_t(X)$, $\overline{R}_t(X)$ and $B_t(X)$ called a *t-lower approximation of X, a t-upper approximation of X* and *a t-boundary of X*, respectively, where

$$\underline{R}_t(X) = \{[x]_t : [x]_t \subseteq X\},$$

$$\overline{R}_t(X) = \{[x]_t : [x]_t \cap X \neq \emptyset\},$$

and

$$B_t(X) = \overline{R}_t(X) - \underline{R}_t(X).$$

Intuitively speaking, a t-lower approximation of X is the collection of all elements of the universe which can be classified by an agent t with full certainty, as elements of X, using her knowledge base $\mathcal{E}_t$. A t-upper approximation of X is the collection of all objects from the universe U which can be possibly classified, by an agent t, as elements of X, using her knowledge base $\mathcal{E}_t$. Finally, a boundary of X is in a sense an undecidable area of the universe, that is, none of the objects belonging to $B_R(X)$ can be classified with certainty by an agent t into X or $-X$ as far as her knowledge base $\mathcal{E}_t$ is concerned.

We say that a concept X is *t-definable* if $\underline{R}_t(X) = \overline{R}_t(X)$. Clearly, if X is t-defined, then $\underline{R}_t(X) = X$, $\overline{R}_t(X) = X$ and $B_t(X) = \emptyset$. If for $X \subseteq U$, $\underline{R}_t(X) \neq \overline{R}_t(X)$, then X is said to be *t-rough*.

Example 2. Suppose that we have three groups of agents s, t and w and their knowledge base about the universe $U = \{o_1, o_2, o_3, o_4, o_5\}$ is as follows: $\mathcal{E}_s = \{\{o_1, o_2\}, \{o_3\}, \{o_4, o_5\}\}$, $\mathcal{E}_t = \{\{o_1\}, \{o_2, o_3\}, \{o_4\}, \{o_5\}\}$ and $\mathcal{E}_w = \{\{o_1, o_2\}, \{o_3, o_4\}, \{o_5\}\}$. Take as X the set $\{o_1, o_2\}$. Then $\underline{R}_s(X) = \underline{R}_w(X) = X$ and $\underline{R}_t(X) = \{o_1\}$, $\overline{R}_t(X) = \{o_1, o_2, o_3\}$, and $B_t(X) = \{o_2, o_3\}$. Hence X is s- and w-definable, and X is a t-rough set.

□

Now, we list selected properties of lower and upper approximations and the boundary of any concept. Proofs can be found in (Rauszer, to appear).

The first lemma says that any lower approximation behaves as an interior operation in a topological space and any upper approximation has the same properties as a closure operation in a topological space.

In particular we have:

Lemma 3.1 *Let $\mathcal{E}_t$ be a knowledge base of an agent t and let $X \subseteq U$. Then*

1. $(U, \underline{R}_t)$ *is a topological space, where $\underline{R}_t$ is an interior operation on U, i.e., for every $X \subseteq U$ and $Y \subseteq U$*
   - $\underline{R}_t(X) \subseteq X$,
   - $\underline{R}_t(X) \subseteq \underline{R}_t(\underline{R}_t(X))$,
   - $\underline{R}_t(X) \cap \underline{R}_t(Y) = \underline{R}_t(X \cap Y)$,
   - $\underline{R}_t(U) = U$.
2. $\mathcal{E}_t$ *is a subbasis of the topological space $(U, \underline{R}_t)$.*
3. $\underline{R}_t(X) = -\overline{R}_t(-X)$, $\quad \overline{R}_t(X) = -\underline{R}_t(-X)$.
4. $\underline{R}_t(\overline{R}_t(X)) = \overline{R}_t(X)$ $\quad \overline{R}_t(\underline{R}_t(X)) = \underline{R}_t(X)$.
5. X *is t-definable if and only if $-X$ is t-definable.*
6. *Let X be t-definable. $X \subseteq Y$ if and only if $X \subseteq \underline{R}_t(Y)$.*
7. *Let Y be t-definable. $X \subseteq Y$ if and only if $\overline{R}_t(X) \subseteq Y$.*

□

Lemma 3.2 *Let T be a family of groups of agents, and let for every $t \in T$, $\mathcal{E}_t$ be a knowledge base of t. The following conditions are true:*

1. $\mathcal{E}_s \prec \mathcal{E}_t$ *if and only if $\underline{R}_t(X) \subseteq \underline{R}_s(X)$.*
2. $\mathcal{E}_s \prec \mathcal{E}_t$ *if and only if $\underline{R}_s(\underline{R}_t(X)) = \underline{R}_t(X)$.*
3. $\mathcal{E}_s \prec \mathcal{E}_t$ *if and only if $\overline{R}_s(X) \subseteq \overline{R}_t(X)$.*
4. $\mathcal{E}_s \prec \mathcal{E}_t$ *if and only if $\overline{R}_s(\overline{R}_t(X)) = \overline{R}_t(X)$.*
5. $\mathcal{E}_s \prec \mathcal{E}_t$ *if and only if $B_s(X) \subseteq B_t(X)$.*
6. $\mathcal{E}_s \prec \mathcal{E}_t$ *implies $\underline{R}_t(\underline{R}_s(X)) = \underline{R}_s(\underline{R}_t(X))$.*

□

Lemma 3.3 *Let T be a family of groups of agents such that $\mathcal{E} = \{\mathcal{E}_t\}_{t \in T}$ is a lattice of partitions. For every set $X \subseteq U$ the following holds:*

1. $\underline{R}_t(X) \cup \underline{R}_s(X) \subseteq \underline{R}_{t \vee s}(X)$.
2. $\underline{R}_{s \wedge t}(X) \subseteq \underline{R}_t(X) \cap \underline{R}_s(X)$.

3. $\underline{R}_t(\underline{R}_s(X)) \subseteq \underline{R}_t(X)$.

4. $\overline{R}_{s \vee t}(X) \subseteq \overline{R}_t(X) \cap \overline{R}_s(X)$.

5. $\overline{R}_t(X) \cup \overline{R}_s(X) \subseteq \overline{R}_{s \wedge t}(X)$.

6. $\overline{R}_s(X) \subseteq \overline{R}_s(\overline{R}_t(X))$.

7. $B_{s \vee t}(X) \subseteq B_s(X) \cap B_t(X)$.

8. $B_s(X) \cup B_t(X) \subseteq B_{s \wedge t}(X)$.

9. $B_t(X) = B_t(-X)$.

10. $\underline{R}_t(B_t(X)) = B_t(X)$.

11. $B_t(X) = \emptyset$ if and only if $\underline{R}_t(X) = X$. $\qquad\qquad\qquad\square$

Let $\mathcal{E}_t$ be a knowledge base of an agent t and let $X \subseteq U$. Recall that a t-lower approximation $\underline{R}_t(X)$ is meant as the set of all objects which for an agent t, according to her knowledge base $\mathcal{E}_t$ have, without any doubt the property given by the concept X. This enables us to interpret a t-lower approximation of X as *knowledge of an agent t about X*. Thus, the set $\underline{R}_t(X)$ might be viewed as the set of all objects which according to the knowledge of an agent t have the property described by the concept X. In other words, the set $\underline{R}_t(X)$ may be treated as the set of all positive examples of X (for an agent t) or the set of all objects which (for an agent t) belong to X. Hence, if an agent t is able to decide, according to her knowledge base $\mathcal{E}_t$, whether a given object x has the property described by the concept X, that is, she knows with certainty whether x belongs to X or to $-X$, then $\underline{R}_t(X) = X$, which means that X is t-defined. In that case the undecidable area about the concept X, that is, the t-boundary of X is empty. If $\underline{R}_t(X) \neq X$, that is, X is t-rough, then the t-lower aproximation of X might be viewed as the set of all objects which are known by an agent t as objects belonging to X. In that case, a t-upper approximation $\overline{R}_t(X) \neq \emptyset$ and as a consequense $B_t(X) \neq \emptyset$, which means that there is at least one object such that the agent t, according to her knowledge base $\mathcal{E}_t$, is not able to justify whether it has the property X or not.

Thus, the equality:

$$\underline{R}_s \underline{R}_t(X) = \underline{R}_t(X).$$

may be read:

an agent s knows what an agent t knows about X.

Thus, immediately from Lemma 2.3 we have

Example 4. Take the information system $S = (U, \mathbf{A})$ from the previous example and let $X = \{o_2, o_5\}$. Then the b-lower approximation of X is $\underline{ind}(b)(X) = \{o_5\}$, and the b-upper aproximation of X is $\overline{ind}(b)(X) = \{o_2, o_3, o_5\}$. Thus, the set $\{o_2, o_5\}$ is b-rough in S. Notice that the set $\{o_2, o_5\}$ is $\mathbf{A}$-defined in $S = (U, \mathbf{A})$. $\qquad\square$

It turns out that every knowledge base may be represented as an information system, that is, in the form of an attribute-value table.

To this end let $\mathcal{E}_t$ be a knowledge base about U. It is not difficult to construct an information system $S_t = (U, \mathbf{A}_t)$ such that $\mathcal{E}_t = \mathcal{E}_{\mathbf{A}_t}$. If for some information system $S_t = (U, \mathbf{A}_t)$, $\mathcal{E}_t = \mathcal{E}_{\mathbf{A}_t}$ then we call $S_t = (U, \mathbf{A}_t)$ a *knowledge representation system (k.r.s.)* for $\mathcal{E}_t$.

Example 5. Suppose that $\mathcal{E}_s = \{\{o_1, o_2\}, \{o_3\}, \{o_4, o_5\}\}$, $\mathcal{E}_t = \{\{o_1\}, \{o_2, o_3\}$, $\{o_4\}, \{o_5\}\}$, and $\mathcal{E}_w = \{\{o_1\}, \{o_2, o_3, o_4\}, \{o_5\}\}$ are partitions of U.

Notice that each $\mathcal{E}_u$, where $u \in \{s, t, w\}$ may be represented by $\mathcal{S}_u$ depicted below:

$\mathcal{S}_t$:

U	a	b	c
o_1	1	2	3
o_2	0	1	3
o_3	0	1	3
o_4	2	1	2
o_5	1	3	2

$\mathcal{S}_s$:

U	c	d	e
o_1	3	0	2
o_2	3	0	2
o_3	3	1	0
o_4	2	4	1
o_5	2	4	1

$\mathcal{S}_w$:

U	f	g
o_1	1	1
o_2	0	0
o_3	0	0
o_4	0	0
o_5	2	2

Observe that the strong distributed knowledge base

$$\mathcal{E}_{s \vee t \vee w} = \{\{o_1\}, \{o_2\}, \{o_3\}, \{o_4\}, \{o_5\}\}$$

and the weak distributed knowledge base

$$\mathcal{E}_{s \wedge t \wedge w} = \{o_1, o_2, o_3, o_4, o_5\}$$

of $\mathcal{E}_s, \mathcal{E}_t$ and $\mathcal{E}_w$ may be represented by $\mathcal{S}_{s \vee t \vee w}$ and $\mathcal{S}_{s \wedge t \wedge w}$, respectively, as follows:

$\mathcal{S}_{s \vee t \vee w}$:

U	a	b	c	d	e	f	g
o_1	1	2	3	0	2	1	1
o_2	0	1	3	0	2	0	0
o_3	0	1	3	1	0	0	0
o_4	2	1	2	4	1	0	0
o_5	1	3	2	4	1	2	2

and

$\mathcal{S}_{s \wedge t \wedge w}$:

U	k
o_1	3
o_2	3
o_3	3
o_4	3
o_5	3

□

From now on the notion knowledge representation system will be used synonymous with information system.

Finally, it seems worthwhile to mention that for the given knowledge representation system $\mathcal{S} = (U, \mathbf{A})$ and $X \subseteq U$, there are several efficient methods for computing sets $\underline{\mathbf{A}}(X)$, $\overline{\mathbf{A}}(X)$ and the boundary of X. One of them is presented in [SR] and says that the upper bound for the time complexity for computing $\underline{\mathbf{A}}(X)$, $\overline{\mathbf{A}}(X)$ and the boundary of X is of order n^2, where n is the number of objects in U.

4 Rough Logic

4.1 Syntax

In this section we are going to define a deductive system which may be served as a logical tool in the investigations of multi-agents systems.

We call this logic *rough logic for multi-agents systems* and denote by R-logic, where R refers to rough.

Let T be a family of groups of agents which perceive the same reality U. Moreover, assume that if $s, t \in T$, then the group st has strong and weak distributed knowledge about U.

Let $\mathbf{A}$ be a finite set of attributes such that $2^{|T|} \leq |\mathbf{A}|$. We associate with every $a \in \mathbf{A}$ a non-empty set V_a, such that $2 \leq |V_a|$ and let $V = \bigcup_{a \in \mathbf{A}} V_a$. Every set V_a may be treated as a domain of an attribute $a \in \mathbf{A}$.

Now, we associate with $\mathbf{A}$ and T a language $\mathcal{L}_{\mathbf{A},T}$ called *the formal language of R-logic*. All elements of $\mathbf{A}$, T and V are treated as constants of $\mathcal{L}_{\mathbf{A},T}$. Elements of $\mathbf{A}$ are called *attribute constants* and denoted by $a, b, c, \ldots$, elements of T are named *agent constants* and denoted $s, t, \ldots$, and elements of V are called *attribute-value constants* and denoted by $v, u \ldots$.

The language $\mathcal{L}_{\mathbf{A},T}$ consists of two levels. The expressions of these levels are called *1-st kind formulae* and *2-nd kind formulae*, respectively. Intuitively, 1-st kind formulae describe relations between knowledge bases understood as a partition of the universe U, whereas 2-nd kind formulae express certain facts about sets of objects of U or approximations of these sets.

To give a formal definition of the sets of formulae, we define first terms of $\mathcal{L}_{\mathbf{A},T}$.

Terms are built up from agent constants, two constants $\mathbf{0}$ and $\mathbf{1}$, and operations: $\vee$ and $\wedge$. More precisely, the *set of all terms* is defined to be the least set $\mathcal{T}$ with the following three properties:

- $\mathbf{0}$ and $\mathbf{1}$ are in $\mathcal{T}$.
- all agent constants are in $\mathcal{T}$,
- $s \vee t$, $s \wedge t$ are in $\mathcal{T}$, whenever s, t are terms.

Formulae of the 1-st kind are built up from terms and a unary operation $\Rightarrow$. In particular, the set F_1 of all *formulae of the 1-st kind* is the smallest set such that

- If $s, t \in \mathcal{T}$, then $s \Rightarrow t \in F_1$.

Intuitively, 1-st kind formulae express a hierarchy between agents, that is determined by their knowledge base. The formula $s \Rightarrow t$ may be interpreted in the following way: a sharpness of recognizing features of objects by an agent s is weaker than that of an agent t.

The set F_2 of all *formulae of the 2-nd kind* is the smallest set containing all atomic formulae which are of the form (a, v), where $a \in \mathbf{A}$ and $v \in V$; it is closed with respect to the propositional connectives $\vee$, $\wedge$, $\rightarrow$, $\leftrightarrow$, $\neg$ and a family $\{I_t\}_{t \in T}$ of modal connectives, that is, the following conditions are satisfied for F_2:

- $(a, v) \in F_2$, for every $a \in \mathbf{A}$ and $v \in V$.
- $\phi \vee \psi$, $\phi \wedge \psi$, $\phi \rightarrow \psi$, $\phi \leftrightarrow \psi$, and $\neg\phi$ are in F_2, whenever ϕ and ψ are in F_2.
- For every $t \in T$, $I_t\phi \in F_2$, whenever ϕ is in F_2.

The axioms for the R-logic consist of three groups: one for 1-st kind formulae, one for 2-nd kind formulae, and one separate group. Specific axioms express characteristic properties of information systems.

The axioms for 1-st kind formulae are:

(t_1) $\qquad t \Rightarrow t$,

(t_2) $\qquad s \Rightarrow s \vee t \qquad t \Rightarrow s \vee t$,

(t_3) $\qquad s \wedge t \Rightarrow s \qquad s \wedge t \Rightarrow t$,

where s and t are any terms.

Now we list axioms for 2-nd kind formulae:

(c) $\qquad$ All axioms for classical logic,

(i_1) $\qquad I_t\phi \rightarrow \phi$,

(i_2) $\qquad I_t\phi \rightarrow I_t I_t\phi$,

(i_3) $\qquad I_t(\phi \wedge \psi) \leftrightarrow (I_t\phi \wedge I_t\psi)$,

(i_4) $\qquad I_t(\phi \vee -\phi)$,

where ϕ is any formula of the second kind, and I_t is any modal operator.

The specific axioms of R-logic are as follows:

1. $(a, v) \wedge (a, u) \leftrightarrow \perp$ for any $a \in \mathbf{A}$, $v, u \in V_a$ and $v \neq u$.
2. $\bigvee_{v \in V_a}(a, v) \leftrightarrow \top$, for every $a \in \mathbf{A}$,
3. $\neg(a, v) \leftrightarrow \bigvee\{(a, u) : u \in V_a, u \neq v\}$, for every $a \in \mathbf{A}$,

where $\phi \vee -\phi =_{df} \top$, $\phi \wedge -\phi =_{df} \perp$ and $\bigvee \phi$ means a finite disjunction.

Specific axioms characterize our notion of knowledge representation system. Observe that the first axiom follows from the assumption that each object can have exactly one value for each attribute. Axiom (2) follows from the assumption that each object in any knowledge representation system S has a value with respect to every attribute. Hence, the description of objects is complete up to a given set of attributes. In other words, for every $a \in \mathbf{A}$ and every object x, the entry in the row x and the column a (in S viewed as a table) is nonempty.

The third axiom allows us to figure out negation in such a way that instead of saying that an object does not possess a given property we can say that it has one of the remaining properties. For example, instead of saying that something is not blue we can say it is either red or green or yellow, etc.

We say that a formula ϕ is *derivable in R-logic from a set of formulae F*, denoted by $F \vdash \phi$, provided it can be concluded from F by means of the above axioms and the following rules: modus ponens, and for every term s, t and w

$$\frac{\phi}{I_s\phi},$$

$$\frac{s \Rightarrow t}{I_s \phi \rightarrow I_t \phi} \; ,$$

$$\frac{s \Rightarrow w \quad t \Rightarrow w}{s \vee t \Rightarrow w} \qquad \frac{w \Rightarrow s \quad w \Rightarrow t}{w \Rightarrow s \wedge t} \; ,$$

$$\frac{s \Rightarrow t \quad t \Rightarrow w}{s \Rightarrow w} \; .$$

If ϕ is derivable in R-logic from the empty set, then we write $\vdash \phi$ and say ϕ *is derivable*. Clearly, all classical tautologies are derivable. Also $\vdash \top$ and $\nvdash \bot$.

It is easy to prove

Lemma 4.1 *The following are derivable formulae in R-logic:*

1. $(I_s \phi \vee I_t \phi) \rightarrow I_{s \vee t} \phi$,
2. $I_{s \wedge t} \phi \rightarrow (I_s \phi \wedge I_t \phi)$,
3. $I_s I_t \phi \rightarrow I_s \phi$,
4. $I_s I_t \phi \rightarrow I_t \phi$. □

In the standard way we may prove the Deduction Theorem:

Theorem 4.2 *Let F be a set of formulae of the 2-nd kind. For every formula ϕ and ψ of the 2-nd kind and every modal operator I_s,*

$$F \vdash I_s \phi \rightarrow \psi \text{ if and only if } F \cup \{\phi\} \vdash \psi.$$

□

Consider now the set of all terms $\mathcal{T}$ as an abstract algebra $(\mathcal{T}, \vee, \wedge, 0, 1)$. Put

$$s \sim t \text{ if and only if } \vdash s \Rightarrow t \text{ and } \vdash t \Rightarrow s.$$

One can prove that $\sim$ is a congruence relation with respect to $\vee$ and $\wedge$. The algebra $(\mathcal{T}/\sim, \vee, \wedge, 0, 1)$ will be called the *algebra of terms of R-logic*. For every term t, $\|t\|$ denotes an element in the algebra of terms.

It can be proved:

Lemma 4.3 *The algebra of terms is a lattice. Moreover,*

1. *$\|s \vee t\| = \|s\| \vee \|t\|$,*

2. *$\|s \wedge t\| = \|s\| \wedge \|t\|$, where $\vee$ and $\wedge$ on the left hand side of the equality are logical connectives and $\vee$ and $\wedge$ on the right hand side of the equality are lattice operations.*

3. *The relation $\leq$ is defined as follows: $\|s\| \leq \|t\|$ if and only if $\vdash s \Rightarrow t$ is an ordering relation on $\mathcal{T}/\sim$, where $s \Rightarrow t$ is a formula of the 1-st kind from $\mathcal{L}_{\mathbf{A},T}$.*

□

Consider the set of all formulae F_2 of the 2-nd kind as an abstract algebra, that is, as an algebra

$$(F_2, \vee, \wedge, \rightarrow, \neg, \{I_t\}_{t\in T})$$

with binary operations $\vee, \wedge, \rightarrow$, unary operation $\neg$, and modal operators I_t, where $t \in T$.

The relation $\approx$ is defined in F as follows:

$$\phi \approx \psi \text{ if and only if } \vdash \phi \leftrightarrow \psi.$$

The relation $\approx$ is a congruence with respect to the operations $\vee, \wedge, \rightarrow, \neg$ and the modal operators I_t, where t is a term. Thus the set $F_2/_\approx$ can be conceived as an abstract algebra $\mathcal{A} = (F_2/_\approx, \vee, \wedge, \rightarrow, \neg, \{I_t\}_{t\in T})$, called the *Lindenbaum algebra of R-logic*. Let for every formula ϕ, $\|\phi\|$ denote an element of $F_2/_\approx$. In the standard way one can prove the following theorem:

Theorem 4.4 *For every modal operator I_t, the I_t-reduct of the Lindenbaum algebra of R-logic $\mathcal{A}$, that is, the algebra $(F_2/_\approx, \vee, \wedge, \rightarrow, \neg, I_t)$ is a topological Boolean algebra.*

Moreover,

1. $\|\phi\| = 1$ *if and only if ϕ is derivable, where $1 = \|\top\|$ is the unit element in $\mathcal{A}$.*
2. $\|\phi \circ \psi\| = \|\phi\| \circ \|\psi\|$,
3. $\| \circ \phi\| = \circ\|\phi\|$,

where $\circ$ on the left hand side of the equality means a logical connective and $\circ$ on the right hand side of the equality means one of the operations in the Lindenbaum algebra of R-logic.

$\square$

4.2   Semantics

Intuitively speaking, formulae of the 1-st kind are meant to describe a hierarchy between groups of agents and formulae of the 2-nd kind are meant as descriptions of subsets of objects obeying properties expressed by these formulae. For instance, a natural interpretation of an atomic formula (a, v) is the set of all objects having value v for the attribute a. Hence, a natural interpretation of a formula of the form $I_t(a, v)$ is a t-lower approximation of the set of all objects having the property v for the attribute a.

Let $\mathcal{L}_{\mathbf{A},T}$ be the language of R-logic described above determined by $\mathbf{A}$ and T. Let U be a non-empty set such that $|U| \geq 2^T$. Let for every term t, $\mathcal{E}_t$ be a partition of U conceived as a knowledge base of a group of agents t and let $\mathcal{E} = \{\mathcal{E}_t\}_{t\in T}$.

Lemma 4.5 . *The family of all partitions determined by terms, that is, the family $\mathcal{E} = \{\mathcal{E}_t\}_{t \in T}$ is a lattice of partitions.*

Proof: By Lemma 5.3 we know that the algebra of terms of R-logic is a lattice. Put, for every $s, t \in T$,

$$\mathcal{E}_t \prec \mathcal{E}_s \text{ if and only if } \|s\| \leq \|t\|,$$

where the relation $\leq$ is defined as in Lemma 5.3

It is not difficult to show that the relation $\prec$ is an ordering relation on $\mathcal{E}$. To prove that $\mathcal{E}$ is a lattice it suffices to assert that for all $\mathcal{E}_s, \mathcal{E}_t$, the infimum and the supremum exist in $\mathcal{E}$. We will only show that $inf\{\mathcal{E}_s, \mathcal{E}_t\}$ exists in $\mathcal{E}$ (for the supremum the proof is similar). Notice, that for any terms s, t, $t \lor s$ is also a term, hence, $\mathcal{E}_{s \lor t} \in \mathcal{E}$. We prove now, that $\mathcal{E}_{s \lor t}$ is upper lower bound of $\mathcal{E}_s$ and $\mathcal{E}_t$ in $\mathcal{E}$. By Lemma 5.3 (3), axiom (t_2) and the definition of $\prec$ we have that $\mathcal{E}_{s \lor t} \prec \mathcal{E}_s$ and $\mathcal{E}_{s \lor t} \prec \mathcal{E}_t$. Now, let $\mathcal{E}_z \prec \mathcal{E}_s$ and $\mathcal{E}_z \prec \mathcal{E}_t$. Then $\vdash x \Rightarrow z$ and $\vdash t \Rightarrow z$, and by the corresponding inference rule we have that $\mathcal{E}_z \prec \mathcal{E}_{s \lor t}$, which was to be shown.

$\square$

Now we are going to define semantics for R-logic. Let $\mathcal{L}_{\mathbf{A},T}$ be the formal language of R-logic defined in the previous subsection. Assume that a universe of discourse U, $|U| \geq 2^T$, is given and let for every $t \in T$, $\mathcal{E}_t$ be a partition of U. Let $\mathcal{S} = (U, \mathbf{A})$ be an information system such that $|U| \geq 2^T$, $|A| \geq 2^T$ and a domain of each attribute contains at least two elements, that is, $|V_a| \geq 2$ for every $a \in \mathbf{A}$. $\mathcal{S} = (U, \mathbf{A})$ is said to be *an information system associated with* $\mathcal{L}_{\mathbf{A},T}$ provided that for every $t \in T$ there is a subset $\mathbf{A}_t$ of $\mathbf{A}$ such that $\mathcal{E}_t = \mathcal{E}_{\mathbf{A}_t}$.

Observe, that for a given language $\mathcal{L}_{\mathbf{A},T}$, an information system associated with $\mathcal{L}_{\mathbf{A},T}$ may be constructed as follows: By the assumption, the powerset of attributes $\mathbf{A}$ of $\mathcal{S} = (U, \mathbf{A})$ is greater than or equal to 2^T, and the power of each attribute domain is at least two. Now, for every $\mathcal{E}_t$, where t is an agent constant, take as $\mathbf{A}_t$ all mappings $a_t : U \to V_{a_t}$, $a_t \in \mathbf{A}$, such that for every $[x]_t \in \mathcal{E}_t$, there is a $v \in V_{a_t}$ for which $\{o \in U : a_t(o) = v\} = [x]_t$. Clearly, for such defined $\mathbf{A}_t$, $\mathcal{E}_t = \mathcal{E}_{\mathbf{A}_t}$.

If for all $\mathcal{E}_t$, where t is an agent constant, $\mathcal{E}_{\mathbf{A}_t}$ is constructed, then it is also constructed for $\mathcal{E}_{s \lor t}$, where s, t are agent constants. Indeed, if s, t are agent constants, then by Lemma 5.5 the partition $\mathcal{E}_{s \lor t}$ associated to term $s \lor t$ is equal to $\mathcal{E}_s \land \mathcal{E}_t$. Hence, as $\mathbf{A}_{s \lor t}$ take $\mathbf{A}_s \cup \mathbf{A}_t$.

Suppose we have defined $\mathbf{A}_z$ for all agent constants z. Thus, $\mathbf{A}_z$ is also defined for all terms z of the form $s \lor t$, where s, t are agent constants. If $\mathcal{E}_{s \land t}$ is different from all $\mathcal{E}_{\mathbf{A}_z}$ which have been constructed, then take as $\mathbf{A}_{s \land t}$ the mapping $a_{s \land t}$, such that for every $[x]_{s \land t} \in \mathcal{E}_{s \land t}$ there is an $v \in V_{s \land t}$ such that $\{o \in U : a_{s \land t}(o) = v\} = [x]_{s \land t}$, where $a_{s \land t}$ is the first unused attribute in $\mathbf{A}$.

Finally, as the required information system take $(U, \bigcup_{t \in T} \mathbf{A}_t)$.

Now, consider the pair $(\mathcal{S}, \mathcal{E})$, where $\mathcal{S} = (U, \mathbf{A})$ is an information system associated with $\mathcal{L}_{\mathbf{A},T}$ and $\mathcal{E} = \{\mathcal{E}_t\}_{t \in T}$ is the family of all partitions determined by terms. Recall, that $\mathcal{E}$ is a lattice of partitions.

We say that 1-st kind formula of the form $s \Rightarrow t$ is *true in* $(S, \mathcal{E})$, denoted by $\models s \Rightarrow t$, if $\mathcal{E}_t \prec \mathcal{E}_s$ holds in the lattice $\mathcal{E}$.

Let ϕ be a formula of the 2-st kind from $\mathcal{L}_{\mathbf{A},T}$. We say that an object $x \in U$ *satisfies a formula* ϕ in $(S, \mathcal{E})$, denoted by $x \models_{(S,\mathcal{E})} \phi$ (or short: $x \models \phi$ if the information system S follows from the context) if and only if the following conditions are satisfied:

1. $x \models (a, v)$ if and only if $a(x) = v$,
2. $x \models \neg\phi$ if and only if $x \not\models \phi$,
3. $x \models \phi \vee \psi$ if and only if $x \models \phi$ or $x \models \psi$,
4. $x \models \phi \wedge \psi$ if and only if $x \models \phi$ and $x \models \psi$,
5. $x \models \phi \rightarrow \psi$ if and only if $x \models \neg\phi \vee \psi$,
6. $x \models \phi \leftrightarrow \psi$ if and only if $x \models \phi \rightarrow \psi$ and $x \models \psi \rightarrow \phi$,
7. For every modal operator I_t, $x \models I_t\phi$ if and only if for all $x_i \in U$ if $x_i \in [x]_t$ then $x_i \models \phi$.

Let $S = (U, \mathbf{A})$ be an information system associated with $\mathcal{L}_{\mathbf{A},T}$. For any formula ϕ of the 2-nd kind the set $|\phi|_S$, defined by

$$|\phi|_S = \{x \in U : x \models \phi\}$$

will be called a *meaning of the formula* ϕ in S. As before we will omit the subscript S, if the information system S follows from the context.

Notice, (cf. Lemma 3.3 (Rauszer 1992a)) that for every meaning $|\phi|_{(U,\mathbf{A})}$ of a satisfiable formula ϕ which is a conjunction of atomic formulae of the form (a, v), $a \in \mathbf{A}$, $v \in V_a$ there is a basic concept $[x]$ such that $|\phi|_{(U,\mathbf{A})} = [x]$, and vice versa, for every basic concept $[x]$ there is a formula ϕ such that $[x] = |\phi|_{(U,\mathbf{A})}$.

Now, we list some basic properties of the meaning of formulae. The simple proof of the lemma given below is left to the reader.

Lemma 4.6 *For every atomic formula (a, v) and any formulae ϕ and ψ of the 2-nd kind the following conditions are true:*

1. $|(a, v)| = \{x \in U : a(x) = v\}$.
2. $|\neg\phi| = -|\phi|$.
3. $|\phi \vee \psi| = |\phi| \cup |\psi|$.
4. $|\phi \wedge \psi| = |\phi| \cap |\psi|$.
5. $|\phi \rightarrow \psi| = -|\phi| \cup |\psi|$.
6. $|\phi \leftrightarrow \psi| = |\phi \rightarrow \psi| \cap |\psi \rightarrow \phi|$.
7. $|I_t\phi| = \bigcup_{[x]_t \subseteq |\phi|} [x]_t$, for every $t \in T$. $\qquad \square$

Let $S = (U, \mathbf{A})$ be an information system associated to $\mathcal{L}_{\mathbf{A},T}$. We say that a formula ϕ of the 2-nd kind is *valid in* $(S, \mathcal{E})$, denoted by $\models \phi$, if $|\phi|_S = U$ for any meaning of ϕ. If any formula ϕ (of the 1-st or of the 2-nd kind) is valid in $(S, \mathcal{E})$, then we call the pair $(S, \mathcal{E})$ a *model for* ϕ. If $(S, \mathcal{E})$ is a model for every formula from a set F, then we say that the pair $S = (U, \mathbf{A})$ is a *model for* F. Finally, we say that F *implies* ϕ, denoted by $F \models \phi$, if from the fact that $(S, \mathcal{E})$ is a model of F follows that $(S, \mathcal{E})$ is a model of ϕ.

The next theorem shows that our axiomatization is sound.

Theorem 4.7 *(soundness) Let F be a set of formulae. If $F \vdash \phi$ then $F \models \phi$.*

$\square$

Suppose now, an information system $S = (U, \mathbf{A})$ associated with $\mathcal{L}_{\mathbf{A},T}$ is given and let for every term t, $\mathcal{E}_t$ be a corresponding partition of U. Recall that the family $\mathcal{E} = \{\mathcal{E}_t\}_{t \in T}$ is a lattice of partitions.

Lemma 4.8 *There is an isomorphism of the algebra of terms of R-logic on the lattice of partition of $\mathcal{E}$.*

Moreover,

$$\|s\| \leq \|t\| \text{ if and only if } \mathcal{E}_t \prec \mathcal{E}_s.$$

Proof: It is easy to show that the mapping h such that for every $t \in T$, $h(\|t\|) = \mathcal{E}_t$ is the required isomorphism. Notice only, that $h(\|s \vee t\|) = \mathcal{E}_{s \vee t} = \mathcal{E}_s \wedge \mathcal{E}_t$. $\square$

Let $S = (U, \mathbf{A})$ be an information system associated with $\mathcal{L}_{\mathbf{A},T}$. Consider the following algebra $\mathcal{P}(U) = (\mathbf{P}(U)/_=, \cup, \cap, \rightarrow, -, \{\mathbf{A}_t\}_{t \in T})$, where $\mathbf{P}(U)$ is the family of all concepts of U, and for any subsets X, Y of U, $X \rightarrow Y$ means the set $-X \cup Y$, and for every $t \in T$ and $X \subseteq U$, $\mathbf{A}_t(X)$ is the $\mathbf{A}_t$-lower approximation of X in S, $\mathbf{A}_t \subseteq \mathbf{A}$. For simplicity, we will denote elements of $\mathcal{P}(U)$ in the same way as subsets of U.

Notice, that every $\mathbf{A}_t$-reduct of the algebra $\mathcal{P}(U)$ is a topological Boolean algebra. The unit element $\mathbf{1}$ of $\mathcal{P}(U)$ is the equivalence class $\{X : X = U\}$. For instance, the meaning of any valid formula belongs to $\mathbf{1}$. Indeed, if ϕ is a formula of the 2-nd kind valid in S, then $|\phi|_S = \{x : x \models \phi\} = U$.

Observe also that

$$|I_t \phi|_S = \mathbf{A}_t |\phi|_S.$$

Indeed, if $S = (U, \mathbf{A})$ is an information system associated to $\mathcal{L}_{\mathbf{A},T}$, then for every partition $\mathcal{E}_t$, $t \in T$ there is a subset $\mathbf{A}_t$ of $\mathbf{A}$ such that $\mathcal{E}_t = \mathcal{E}_{\mathbf{A}_t}$. Moreover, for every basic concept $[x]_{\mathbf{A}_t}$ there is a formula ϕ such that $|\phi|_S = [x]_{\mathbf{A}_t}$. Hence by Lemma 5.6 (7) we conclude the required equality.

Denote by ∇ the set of all meanings of formulae which are provable, that is,

$$\nabla = \{|\phi|_S : \vdash \phi\}.$$

Lemma 4.9 *For every term t, ∇ is an $\mathbf{A}_t$-filter in $\mathcal{P}(U)$.*

Proof: Clearly, ∇ is closed with respect to $\cap$. Notice, that $\mathbf{1}$ belongs to ∇ as for any formula ϕ, $\vdash \phi \vee -\phi$ and $|\phi \vee -\phi|_S = \mathbf{1}$. Now, let $|\phi|_S \in \nabla$ and let $|\phi|_S \subseteq |\psi|_S$. Then $|\phi \rightarrow \psi|_S = \mathbf{1}$ and therefore $|\phi| \rightarrow |\psi| \in \nabla$. Hence, $\vdash \psi$, which proves that $|\psi|_S \in \nabla$. We claim that for every $\mathbf{A}_t$, where t is a term, $\mathbf{A}_t |\phi|_S \in \nabla$ provided $|\phi|_S \in \nabla$ If $|\phi|_S \in \nabla$, then $\vdash \phi$ and by the necessitation rule, for every modal operator I_t we have $\vdash I_t \phi$. Hence, $|I_t \phi|_S = \mathbf{A}_t |\phi|_S$ and we infer that $\underline{\mathbf{A}}_t|\phi|_S \in \nabla$, which ends the proof that ∇ is $\underline{\mathbf{A}}_t$-filter in the algebra $\mathcal{P}(U)$.

$\square$

We prove the following lemma:

Lemma 4.10 *There is one-to-one monomorphism of the Lindenbaum algebra of R-logic $\mathcal{A}$ into $\mathcal{P}(U)$.*

Proof: Let h be a mapping from $F_2/\approx$ into $\mathbf{P}(U)/_=$ such that $h(||\phi||) = |\phi|_S$, where $|\phi|_S$ is the meaning of ϕ in $\mathcal{S}$ and ϕ is a formula of the 2-nd kind. Clearly, h is a monomorphism of $\mathcal{A}$ into $\mathcal{P}(U)$. To show that h is one-to-one suppose that for some formulae ϕ and ψ, $||\phi|| \neq ||\psi||$. Then either $\nvdash \phi \to \psi$ or $\nvdash \psi \to \phi$. If $\nvdash \phi \to \psi$ then $|\phi \to \psi|_S \notin \nabla$. Therefore, $|\phi|_S \nsubseteq |\psi|_S$. Thus $|\phi|_S \neq |\psi|_S$. In the latter case the proof is analogous.

$\square$

Now, we prove that our axiomatization is complete.

Theorem 4.11 *(completeness) Let F be a set of formulae and let ϕ be a formula of $\mathcal{L}_{\mathbf{A},T}$. If $F \models \phi$ then $F \vdash \phi$.*

Proof: Let ϕ be a formula of the 1-st kind. Suppose that ϕ is of the form $s \Rightarrow t$. Let $(\mathcal{S}, \mathcal{E})$ be a model for F. Recall that the family $\mathcal{E} = \{\mathcal{E}_t\}_{t \in T}$ is a lattice of partitions. By the assumption, $s \Rightarrow t$ is true in $(\mathcal{S}, \mathcal{E})$, and we conclude that $\mathcal{E}_t \prec \mathcal{E}_s$. Suppose now, that $F \nvdash s \Rightarrow t$. Then $\nvdash s \Rightarrow t$ and by Lemma 5.3 we infer that the relation $||s|| \leq ||t||$ does not hold in the algebra of terms. This proves, by Lemma 5.8, that $\mathcal{E}_t \nprec \mathcal{E}_s$, is a contradiction.

Assume now, that ϕ is a formula of the 2-nd kind and let $F \nvdash \phi$. We show that there is a model for F which is not a model for ϕ. By the assumption $\nvdash \phi$ and by Theorem 5.4 $||\phi|| \neq 1$, where $\mathbf{1} = ||T||$ is the unit element in the Lindenbaum algebra of R-logic $\mathcal{A}$. Let $(\mathcal{S}, \mathcal{E})$ be a model for F and let h be a monomorhpism from $\mathcal{A}$ into $\mathcal{P}(U)$ defined as in Theorem 5.10, that is, $h(||\phi||) = |\phi|_S$. Clearly, $|\phi|_S \notin \nabla$, where ∇ is the filter defined above. Thus, because of h is one-to-one we infer that $|\phi|_S \neq U$, the unit element in the algebra $\mathcal{P}(U)$, which proves that $(\mathcal{S}, \mathcal{E})$ is not a model for ϕ.

$\square$

For every term t define now new modal connectives C_t and Br_t as follows: for any formula of the 2-nd kind ϕ put:

$$C_t \phi \equiv \neg I_t \neg \phi$$

and

$$Br_t \phi \equiv C_t \phi \wedge \neg I_t \phi.$$

It is easy to check that

$$|C_t \phi|_S = \overline{\mathbf{A}}_t(|\phi|_S),$$

where C_t is the unary connective defined above, and $\overline{A}_t$ is the upper approximation determined by the relation $ind(A_t)$ in the information system $S = (U, A)$ associated with $\mathcal{L}_{A,T}$.

Moreover,

$$|Br_t\phi|_S = B_t(|\phi|_S),$$

where $B_t(|\phi|_S)$ means the t-boundary of $|\phi|_s$ in S determined by $ind(A_t)$.

Theorem 4.12 *Let $S = (U, A)$ be an information system associated with $\mathcal{L}_{A,T}$ and let $\mathcal{E} = \{\mathcal{E}_t\}_{t\in T}$ be a lattice of partitions, where each $\mathcal{E}_t$ is a partition of U. The following conditions are equivalent:*

1. $X \subseteq U$ *is t-definable in S.*
2. *A formula of the form $I_t\phi \leftrightarrow \phi$ is valid in $(S, \mathcal{E})$, where the meaning of ϕ is X, that is, $|\phi|_S = X$.*
3. $Br_t\phi$ *is false, where ϕ is as before.*
4. $|I_t\phi|_S = X$.

Proof: The proof is very simple. We show only that (1) implies (2). If X is t-definable, then $\underline{A}_t(X) = X$, and therefore $X = \bigcup_{[x]_B \subseteq X}[x]_B$. Hence, there is a formula ϕ such that $|\phi|_s t = X$. Notice, $|\phi \rightarrow I_t\phi|_S = -|\phi|_S \cup |I_t\phi|_S = -X \cup \underline{A}_t X = U$, which together with axiom (i_1) proves that a formula of the form $I_t\phi \leftrightarrow \phi$ is true in $(S, \mathcal{E})$. □

Finally we have

Theorem 4.13 *A pair $(S, \mathcal{E})$ is a model for a formula ϕ if and only if there is a subset A_t of A such that the meaning of ϕ in the information system (U, A_t) is equal to U.* □

5 What an Agent Knows about the Knowledge of Another Agent

We might be interested in investigating whether knowledge $\mathcal{E}_s$ can be expressed in terms of knowledge $\mathcal{E}_t$. In other words, we want to know whether properties of objects expressed in terms of attributes A_s can be expressed in terms of attributes A_t.

Let us start with some definitions:

We say that a *set of attributes A_t depends on a set of attributes A_s* in an information system $S = (U, A)$, abbreviated by $A_s \rightarrow A_t$, if objects which are identified by A_s are also identified by A_t. In other words,

$$A_s \rightarrow A_t \text{ if and only if } ind(A_s) \subseteq ind(A_t).$$

Let $\mathcal{L}_{\mathbf{A},T}$ be a formal language of R-logic and let $(\mathcal{S}, \mathcal{E})$ be a model for $\mathcal{L}_{\mathbf{A},T}$. For any $\mathbf{A}_s, \mathbf{A}_t \subseteq \mathbf{A}$ by the st-rule we mean an implication of the form:

$$(a_1, v_1) \wedge \ldots \wedge (a_n, v_n) \rightarrow (b_1, u_1) \wedge \ldots \wedge (b_m, u_m),$$

which is satisfiable by at least one object, where $\mathbf{A}_s = \{a_1, \ldots, a_n\}$ and $\mathbf{A}_t = \{b_1, \ldots, b_m\}$.

Any information system $(U, \mathbf{A}_s \cup \mathbf{A}_t)$ is said to be *consistent*, provided $\mathbf{A}_s \rightarrow \mathbf{A}_t$ holds in $(U, \mathbf{A}_s \cup \mathbf{A}_t)$.

The next theorem follows from results of previous sections:

Theorem 5.1 *Let $\mathcal{S}_{st} = (U, \mathbf{A}_s \cup \mathbf{A}_t)$ be an information system. Then the following conditions are equivalent:*

1. *An agent s knows what an agent t knows about the universe U,*
2. *An information system $\mathcal{S}_{st}$ is consistent,*
3. *The dependency $\mathbf{A}_s \rightarrow \mathbf{A}_t$ holds in $\mathcal{S}$,*
4. *Each st-rule is valid in $\mathcal{S}_{st}$,*
5. $\underline{\mathbf{A}}_s(\underline{\mathbf{A}}_t(X)) = \underline{\mathbf{A}}_t(X)$,
6. *For every $X \subseteq U$, $\underline{\mathbf{A}}_t(X) \subseteq \underline{\mathbf{A}}_s(X)$, $\overline{\mathbf{A}}_s(X) \subseteq \overline{\mathbf{A}}_t(X)$, $B_s(X) \subseteq B_t(X)$,*
7. $\underline{\mathbf{A}}_s([x_1]_t) \cup \ldots \cup \underline{\mathbf{A}}_s([x_k]_t) = U$, *where $\mathcal{E}_{\mathbf{A}_t} = \{[x_1]_t, \ldots, [x_k]_t\}$,*
8. $\mathcal{E}_s \prec \mathcal{E}_t$,
9. *For every $i \leq k$, $\vdash I_s \phi_i \leftrightarrow \phi_i$, where $|\phi_i|_t = [x_i]_t$, and $[x_i]_t$ is a basic concept of an agent t,*
10. $\vdash I_s I_t \phi \leftrightarrow I_t \phi$, *for any formula ϕ.*

Proof: By Lemma 3.4 (1) is equivalent to (5) and (8). Clearly (2) and (3) are equivalent, as well as, (3), (6) and (8). According to the completeness theorem (9) and (8) are equivalent. (5) implies (10) and (10) implies (9).

Now, we show that (7) and (9) are equivalent.

Assume (7). Then for every $i \leq k$, $\underline{\mathbf{A}}_s([x_i]_t) = [x_i]_t$. Otherwise, there is an object o, such that $o \in [x_i]_t$ and $o \notin \underline{\mathbf{A}}_s([x_i]_t)$. Then for some $[x_j]_t$, $i \neq j$ $o \in \underline{\mathbf{A}}_s([x_j]_t) \subseteq [x_j]_t$, and therefore $o \in [x_j]_t$, a contradiction. Thus by the completeness theorem (7) and (9) are equivalent.

We have now shown that all conditions except (4) are equivalent.

Observe, that (4) is equivalent to (8). Indeed, suppose (4). It suffices to prove that for every s-basic concept $[x]_s$ there is a t-basic concept $[x]_t$ such that $[x]_s \subseteq [x]_t$. Take $[x]_s$. Then, there is a formula ϕ which is a conjunction of atomic formulae of the form $(a, v), a \in \mathbf{A}_s$ such that $|\phi|_{(U,\mathbf{A}_s)} = [x]_s$. Now, as the required t-basic concept take $|\phi|_t$ where $\phi \rightarrow \psi$ is a st-rule valid in $\mathcal{S}_{st}$. On the other hand, assume (8) and that there is an st-rule $\phi \rightarrow \psi$ which is not valid in $\mathcal{S}_{st}$. Then by the remark before Lemma 5.6, (Lemma 3.3 in (Rauszer 1992a)) there are basic concepts $[x]_{|\phi|}, [x]_{|\psi|}$ such that $[x]_{|\phi|} \not\subseteq [x]_{|\psi|}$. By (8) and the quoted lemma, there is a t-basic concept $[x]_t$ which contains $[x]_{|\phi|}$. Moreover, there is ψ' such that $\phi \rightarrow \psi'$ is a valid st-rule. A contradiction, which ends the proof of Theorem 6.1.

$\square$

6 Connections with Some Related Results

In the discussion above, $I_t\phi$ might be read *agent t knows* ϕ and $I_{svt}\phi$ might be treated as a strong distributed knowledge of the group of agents st. In other words, the modal connective I_{svt} is interpreted as the syntactical counterpart of a strong distributed knowledge base of the group st.

Observe that for all terms s, t, I_{svt} is complete with respect to axioms for the operator D described in (Halpern and Moses 1992) and called distributed knowledge.

Indeed, for any terms s, t and formulae of the 2-nd kind ϕ, ψ and $\psi_i, i = 1, \ldots, n$ we have:

$$\vdash I_s\phi \to I_{svt}\phi \qquad\qquad \vdash I_t\phi \to I_{svt}\phi\ ,$$

$$I_{svt}\phi \wedge I_{svt}(\phi \to \psi) \to I_{svt}\psi\ ,$$

and

$$\text{if } \vdash \psi_i \wedge \ldots \wedge \psi_n \to \phi \text{ then} \vdash I_1\psi_1 \wedge \ldots \wedge I_n\psi_n \to I_u\phi\ ,$$

where $u = 1 \vee \cdots \vee n$ and $i \leq$ is a term.

In (Grzymala Busse, to appear), (Orlowska 1989), (Rasiowa 1990), knowledge is understood as an ability of agents to describe all positive examples and all negative examples of the concept. More exactly, knowledge of an agent about a concept X is based on the following intuition: If an agent t knows X, then he can decide for any object if it has the property X or the property $-X$. As a consequence, an agent t can decide membership question for the concept X.

In terms of R-logic, knowledge understood in this fashion may be reflected as follows: For every term t and a formula ϕ of the 2-nd kind put

$$K_t\phi =_{def} I_t\phi \vee I_t - \phi,$$

and let $K_t\phi$ be read *an agent t knows* ϕ.

The following properties of modal operators K_t are a simple consequence of the previous lemmas:

Lemma 6.1 *For every terms s, t and formulae of the second kind ϕ, ψ the following holds:*

1. $K_t\phi \to K_t K_t\phi$,
2. $-K_t\phi \to K_t - K_t\phi$,
3. $K_t\phi \wedge K_t\psi \to K_t(\phi \wedge \psi)$,
4. $K_s\phi \vee K_t\phi \to K_{svt}\phi$,
5. $K_{s\wedge t}\phi \to K_s\phi \wedge K_t\phi$.

□

Lemma 6.2 *For every modal operator K_t and formulae ϕ, ψ the following formulae are not provable in R-logic:*

1. $K_t\phi \to \phi$,
2. $(\phi \to \psi \wedge K_t\phi) \to K_t\psi$,
3. $(K_t\phi \wedge K_t(\phi \to \psi)) \to K_t\psi$,
4. $K_t\phi \vee K_t\psi \leftrightarrow K_t(\phi \vee \psi)$,
5. $K_t(\phi \wedge \psi) \to K_t\phi \wedge K_t\psi$.

□

Observe, that conditions (1) and (2) of the above lemma show that certain paradoxes of epistemic logic are eliminated in R-logic with the family of K_t modal connectives. Namely, by (1) *what an agent knows is true* is not longer valid in R-logic. Also the logical omniscience problem (Hintikka 1962) which says *an agent always knows all the consequences of her knowledge* is not valid in our logic.

Moreover, it follows from condition (3) that knowledge of agents, understood as the ability to classify objects as positive and negative instances of a given property, is not closed under modus ponens.

References

Grzymala Busse, J.W.: *LERS- A System for Learning from Examples Based on Rough Sets*, to appear.

Hintikka, J. *Knowledge and Belief.* Chicago: Cornell University Press, 1962.

Halpern J.Y., and Y. Moses: "A guide to completeness and complexity for modal logics of knowledge and belief," in: *Artificial Intelligence*, 54 (1992) 309 - 379.

Minski, M.: "A Framework for representation knowledge," in: P. Winston (ed.), *The Psychology of Computer Vision*, New York: McGraw-Hill (1975) 211-277.

Pawlak, Z.: "Rough sets," in: *International Journal of Computer and Information Sciences*, 11 (1982) 341-346.

Pawlak, Z.: *Rough Sets - Theoretical Aspects of Reasoning about Data.* Kluwer Academic Press, 1992.

Orlowska E.: "Logic for reasoning about knowledge," in: *Zeitschr.f. Math. Logik und Grundlagen d. Math.*, 35 (1989) 559-572.

Rasiowa, H.: "On approximation logics: A survey." in: *Jahrbuch 1990 der Kurt Gödel Gesellschaft*, (1990) 63-87.

Rauszer, C.M.: "Logic for information systems," in: *Fundamenta Informaticae*, vol.16, no 3-4 (1992a) 371-383.

Rauszer, C.M.: "Knowledge representation systems for groups of agents," in: J. Woleński (ed.), *Philosophical Logic in Poland*, Kluwer, to appear.

Rauszer C.M.: "Distributive knowledge representation systems," *ICS Research Report* 22/92, Warsaw: University of Technology, Institute of Computer Science, 1992b.

Skowron, A.: "On topology in information systems," in: *Bulletin of the Polish Academy of Sciences*, vol.36, no 7-8 (1988) 477-479.

Skowron, A., and C.M. Rauszer: "The discernibility matrices and functions in information systems," in: R. Słowiński (ed.), *Decision Support by Experience-Applications of the Rough Sets Theory*, Kluwer Academic Press, 1992.

A Logical Approach to Multi-Sources Reasoning

Laurence Cholvy

ONERA-CERT, 2 avenue Edouard Belin, 31055 Toulouse, France
E-mail: cholvy@dryas.cert.fr

Abstract. This paper presents two logics for reasoning with information from several sources. The main problem is that the information might be contradictory; we show that ordering the different sources according to their reliability is a good way for solving this problem. The two logics which are presented, correspond to two attitudes one can take with respect to such an order.

1 Introduction

In this paper, we focus on the problem of merging autonomous knowledge based systems.

Such a merge is necessary when one wants to access several databases at the same time i.e., when one wants to query several existing databases and obtain answers that combine information from each of them. In this case, the merging may be logical and not necessarily physical. However, the different databases are merged from the point of view of the users who query them and the query evaluator must compute answers with these logically merged data.

A merge is also necessary when one wants to combine the knowledge of different expert systems in order to build an integrated system. For instance, each expert system may be the specialist for a part of a problem and attacking the whole problem requires a merge.

A merge is furthermore required when one wants to combine the beliefs of several agents. For instance, each agent may have some (possibly incomplete) belief about a given situation and one wants to combine these beliefs in order to reason about this situation. A good example is a police investigation that gathers the accounts of the different witnesses. Each witness has some beliefs about the crime: the first one saw a man but could not hear him, the second one heard a woman's voice, the third one saw two people jumping into a car, etc.

Merging several autonomous data/knowledge bases gives rise to different problems. One of them is the problem of global consistency. Even though each separate system may be consistent, the global one obtained by combining them may no longer be consistent.

The problem is then how to deal with this inconsistent set of information. In database terminology, this means: how can we answer queries addressed to the global database if this database is inconsistent? Or, in the language of belief set merging, the question becomes: what can be believed in an inconsistent context?

Simplifying the problem somewhat and assuming that the database can be described by a set of propositional formulae, the problem is: which theorems may be deduced from an inconsistent set obtained by merging different consistent sets?

In Section 2, we describe some work on reasoning with inconsistency. We will first detail approaches which assume that the different pieces of information are equally trusted. Then, we will describe approaches in which additional information is provided that can be used in order to restore consistency. Our solution to the problem of merging incompatible databases belongs to the second kind of approach: we assume that the bases are ordered according to a total order which reflects their reliability. In Section 3, we will show that there are two possible attitudes with respect to such an order. Section 4 and Section 5 will present two logical systems, corresponding to the two attitudes. Links between them are given in Section 6. Relations with the problems of database updates and belief base updates are established in Section 7.

2 Previous Work on Inconsistent Knowledge Bases

2.1 Paraconsistent logics

Classical logic satisfies the principle of "ex falso sequitur quodlibet" that says that everything can be concluded from a contradiction $(A \wedge \neg A) \rightarrow B$. Classical logic collapses under inconsistency: everything can be deduced from an inconsistent database.

Paraconsistent logics are logics that reject the principle of "sequitur quodlibet". We refer to (Besnard 1990) for a survey.

Here, we illustrate a paraconsistent logic C_w, on an example which comes from (Baral et al. 1992).

Example 1. Assume three databases that contain a police inspector's knowledge and two witnesses's accounts: the inspector knows that if a car is black, then it is dark; if it is white, then it is light; nothing can be both light and dark. The first witness, Bill, saw a black car while the second, John, saw two men in a white car. We formalize this information by:

$inspector = \{$ *(1) black $\rightarrow$ dark,   (2) white $\rightarrow$ light,   (3) $\neg$ light $\vee \neg$ dark* $\}$
$Bill =$ $\{$ *(4) black* $\}$
$John =$ $\{$ *(5) white ,   (6) two*$\}$

The combined knowledge base $(inspector \cup Bill \cup John)$ is inconsistent. Using first order logic to reason with this set, we could conclude everything i.e., we could conclude for instance that the car was a limousine. Using C_w, we can conclude $(dark \wedge light)$ i.e., the inspector cannot know the real type of colour. However, we cannot conclude that the car was a limousine (this is indeed what the inspector does in the real life situation: he does not conclude anything when a contradiction arises).

2.2 Maximal consistent sets

In the next approach, the underlying logic is still classical, but a difference between formulae is made outside the logic; at the meta-level an attempt is made to eliminate the inconsistency by determining maximal consistent subsets.

Combining knowledge bases (Baral et al. 1992) addresses the problem of combining the knowledge of different experts.

They assume a set of consistent theories and a set of integrity constraints. Each theory satisfies the integrity constraints. They aim to combine the given set of theories so that the combined theory is also consistent with respect to the integrity constraints and contains as much consistent information as possible.

According to their approach, the combination of several theories may lead to several resulting theories. So they define the combination of theories as a mapping from a set of theories (the given knowledge bases) and a set of integrity constraints into a set of theories. They provide several alternative definitions of combination functions. For instance, function $Comb1$, takes the union of the theories $T1 \cup \ldots \cup T$ and identifies the maximal subsets that are consistent with the integrity constraints IC.

On the previous example (where there is no integrity constraint), we get the following maximal consistent belief sets.

$Comb1\{ \{ inspector, Bill, John \}, \emptyset \} = \{ \{(1), (2), (3), (4), (6)\},$
$\{(1), (2), (3), (5), (6)\},$
$\{(1), (2), (4), (5), (6)\},$
$\{(2), (3), (4), (5), (6)\} \}$

A formula is a theorem of the combined theory iff it is a theorem of all its subsets. So, according to this combination, "there were two men in the car" is deducible from the combined theory, as well as "the car was black or white". However, "the car was black" and "the car was white" are not deducible.

This means that, after combining the accounts of the different witnesses, the inspector can affirm that "there were two men in the car" (since one witness told him and the other denied it) and that "the car was black or white" (since, although the witnesses are contradictory on this point, the first one told him the car white and the other told him the car was black). However, he can not affirm the real colour of the car.

Sure and doubtful information In (Cholvy 1990) and (Bauval and Cholvy 1991), a similar approach is described for a slightly different context: we try to circumscribe the contradiction as precisely as possible and to distinguish between information that is related to the contradiction and information which is not. The terms "sure information" and "doubtful information" are introduced. Again, the main notions here are the maximal consistent subsets of the database (integrity constraints are not treated separately). A modal logic is used to formally describe the meaning of "sure" and "doubtful" information, but the query evaluator is first order.

For instance, "there were two men in the car" is sure information, since it does not give rise to a contradiction, i.e., it belongs to the intersection of all maximal consistent subsets of the global theory. However "the car was white", "the car was black"...are doubtful information: they are true in one maximal consistent subset and false in another one.

2.3 Paraconsistent logics and extra-information

(Besnard 1990) described the logic $V1$, originally developped by Arruda. $V1$ holds the middle between classical logic and a purely inconsistent logic since it contains two kinds of propositional letters: some that behave classically and some that behave paraconsistently. Its axiomatisation is obtained by adding (as an axiom) the ex-falso quodlibet only for propositions that behave classically.

Making a disctinction between classical and non-classical proposition requires extra-information attaching to propositions and this is the responsibility of the user.

Using $V1$ is impossible when one cannot decide which propositions must behave classically. For instance, the inspector from our example cannot decide, before listening to the witnesses, on which notions they will agree.

2.4 Tagged data and ordered data

(Fagin et al. 1983) dealt with a related problem of updating a database (see Section 7 for a comparision). In order to construct the resulting database in a minimal way (a minimal number of formulae must be rejected) and in order to distinguish between formulae, they introduced the notion of tagged sentences: a tag is a number, attached to a sentence of the global database, which expresses a priority. The lower the tag, the higher the priority during the update process. The sentences that have the tag 0 have the highest priority and cannot be rejected during the update.

Let us reiterate the inspector example in the context of (Fagin et al. 1983) and assume that inspector's formulae are tagged 0, Bill's sentence is tagged 1 and John's sentences are tagged 2. Let us assume that the inspector first listens to Bill. So Bill's account provides an update to his own knowledge. Since there is no contradiction, the result is: $\{(1), (2), (3), (4)\}$. Then he listens to John who tells him (5). As a consequence his knowledge base becomes: $\{(1), (2), (3), (4), (5)\}$. Then John adds (6). Since (6) contradicts the current base and since (6) is tagged 2 (the lowest priority), it is not added. So the inspector's knowledge base is: $\{(1), (2), (3), (4), (5)\}$.

In (Gabbay and Hunter 1991) the authors suggested that reasoning with inconsistency can be studied as a Labelled Deductive System, where deduction is done both on labels and on formulae. (Indeed, in the logic they defined, units of information are labelled formulae).

(Dubois et al. 1992) adapted this approach to the problem of merging knowledge bases in the context of a possibilistic logic: every formula is associated with a tag, referring to the sources of the information: in this way both the level of support of the formula by a source and the reliability of the source itself can be handled.

(Cayrol et al. 1992) finally, assume that the existence of an order on propositions (reflecting some kind of preference). They infer different orders on sets of propositions in such a way that the problem of consistency restoration leads to select consistent subsets which are maximal according to this order.

2.5 Prioritized theories

The last part of (Baral et al. 1992) focus on the combination of prioritized theories. They assume that different knowledge bases may have different priorities. They use priority information when combining the bases. For example, if a base with higher priority contradicts another base with a lower priority, the first one takes precedence over the second. They define two algorithms that combine prioritized theories $T1 \prec T2 \ldots \prec Tk$. The first algorithm, called "bottom-up", starts combining $T1$ and $T2$ with a preference for $T2$. The combined theory then is a set of theories $\{T: T$ is a superset of $T2$ and $T \setminus T2$ is a maximal subset of $T1$ such that T is consistent$\}$. The result is then combined with $T3$ and so on until Tk is reached. The result is finally combined with the set of integrity constraints IC (that have the highest priority). The second algorithm, called "top-down", starts by combining Tk and IC. The result is then combined with $Tk - 1$, and so on until $T1$ is reached.

On the inspector example, the inspector may order the different bases, expressing that: $John \prec Bill \prec inspector$. This means that he estimates that he is himself more reliable than Bill who is himself more reliable than John.

Using the bottom-up algorithm, the results are the two following sets: $\{1, 2, 3, 4, 6\}$ and $\{1, 2, 3, 5, 6\}$.

Using the top-down algorithm, the result of the combination of the three bases generates the set: $\{1, 2, 3, 5, 6\}$.

3 Context of Our Work

We think that the idea of ordering the different sources of information provides a good point of departure since the order may reflect the fact that the sources are not equally reliable.

Reconsider our example. Assume that a first witness, Bill, said that he saw a black car, while a second, John, said that he saw two men in a white car. Assume that the inspector himself has some additional information which is true: for instance, he is sure that the crime was committed on a foggy day. Bringing the fog into play, he might assume that John is less reliable than Bill since he was standing too far from the location of the crime and could not see well because of the fog. So, the inspector may trust him less. In this case, the inspector's prioritization will be: *Inspector> Bill> John*.

We assume that different databases are ordered according to a total order $>$. However, there are two possible attitudes with respect to this order relation:

- The "suspicious" attitude: suspect all the information provided by a database if this database contradicts a more trustable database. The suspicious attitude of the inspector would lead him to conclude no more than that the car was black.
- The "trusting" attitude: suspect the information provided by a database that contradicts the information provided by a more reliable database; all other information is kept. According the trusting attitude, the inspector will conclude that they were two men in a black car. Indeed, concerning the colour of the car, he trusts Bill more than John, so he can assume that the colour is black. Concerning the number of persons, John provides new information that does not contradict Bill's account.

In the two next sections, we study logical systems for reasoning according to the suspicious attitude (Section 4) and to the trusting attitude (Section 5).

Notice that in both cases, we do not construct the merged database outside the logic, as opposed to (Fagin et al. 1983) and (Baral et al. 1992). We want a system which hypothetically generates the formulae that can be deduced in the merged database. So our approach belongs to the hypothetical approaches along the lines of (Farinas Del Cerro and Herzig 1986, Farinas and Herzig 1992).

4   Merging Two Databases According to the Suspicious Attitude

We assume that our language L is a finite set of propositional variables: p_1, p_2, ... p_L.

We denote the two databases to be merged 1 and 2, respectively. We assume that they are finite sets of literals (positive or negative propositional variables). We also assume that they are consistent i.e., they do not contain both a literal and its negation. But we do not assume that they are complete.

We define a logic, called $FUSION - S(1,2)$, whose language L' is obtained from L by adding pseudo-modalities i.e., markers on propositional formulae. These pseudo-modalities are:

- $[i_1 i_2]$ where $i_1 \in \{1,2\}$ and $i_2 \in \{1,2\}$ and $(i_1 \neq i_2)$

$[i_1 i_2]F$ (where F is a formula of L), will mean that, when considering the total order on $\{1,2\} : i_1 > i_2$, F is true in the database obtained by virtually merging database i_1 and database i_2.

Notice that the general form of these pseudo modalities allows us to represent the particular case: $[i]F$, $i=1$ or $i=2$, which means that F is true in database i.

4.1 Semantics

The semantics of $FUSION - S(1,2)$ is the following:

An interpretation of $FUSION - S(1,2)$ is the pair: $M = (W, r)$, where:

- W is the finite set of all interpretations of the underlying language L.
- r is a set of four equivalence relations between interpretations of W.
  Each pseudo-modality is associated to a relation. So, if $[O]$ is a pseudo-modality, $R(O)$ denotes the associated equivalence relation and $\overline{R}(O)$ its equivalence class. The equivalence classes are defined by:
  $\overline{R}(i)$ is a non empty subset of $W, i = 1,2$
  $\overline{R}(i_1 i_2) = \overline{R}(i_1) \cap \overline{R}(i_2)$ if not empty, $(i_1 \in \{1,2\}, i_2 \in \{1,2\}, i_1 \neq i_2)$
  $\overline{R}(i_1 i_2) = \overline{R}(i_1)$ else $(i_1 \in \{1,2\}, i_2 \in \{1,2\}, i_1 \neq i_2)$

Definition 1. Satisfaction of formulae.
Let F be a formula of L. Let $F1$ and $F2$ be formulae of L'. Let O be a total order on a subset of $\{1,2\}$. Let $M = (W, r)$ be an interpretation of $FUSION - S(1,2)$ and let $w \in W$.

$FUSION - S(1,2), r, w \models F$ $\iff$ $w \models F$

$FUSION - S(1,2), r, w \models [O]F$ $\iff$ $\forall w' w' \in \overline{R}(O) \Longrightarrow w' \models F$

$FUSION - S(1,2), r, w \models \neg F1$ $\iff$ $not(FUSION - S(1,2), r, w \models F1)$

$FUSION - S(1,2), r, w \models F1 \wedge F2$ $\iff$ $(FUSION - S(1,2), r, w \models F1)$ and $(FUSION - S(1,2), r, w \models F2)$

Definition 2. Valid formulae in $FUSION - S(1,2)$.
Let F be a formula of L'. F is a valid formula (in $FUSION - S(1,2)$) iff
$\forall M = (W, r), \forall w \in W, FUSION - S(1,2), r, w \models F$

We note $FUSION - S(1,2) \models F$, the valid formulae F.

4.2 Axiomatics

Let us first introduce the following formula:

$$INC(1,2) = (\ ([1]p_1 \wedge [2]\neg p_1) \vee ([1]\neg p_1 \wedge [2]p_1) \vee \ldots \vee$$
$$([1]p_L \wedge [2]\neg p_L) \vee ([1]\neg p_L \wedge [2]p_L))$$

$INC(1,2)$ expresses that there is a literal whose valuation is not the same in database 1 and database 2. So, $INC(1,2)$ is true when the two databases are contradictory.
Axioms of $FUSION - S(1,2)$ are:

- $(A0)$ the axioms of propositional logic
- $(A1)$ $[O]\neg F \rightarrow \neg[O]F$ if F is a formula of L

- (A2) $[O]F \wedge [O](F \rightarrow G) \rightarrow [O]G$ if F is a formula of L
- (A3) $INC(1,2) \rightarrow ([i_1 i_2]l \longleftrightarrow [i_1]l)$ if l is a literal
- (A4) $\neg INC(1,2) \rightarrow ([i_1 i_2]l \longleftrightarrow ([i_1]l \vee [i_2]l))$ if l is a literal

Inference rules of $FUSION - S$ are:

- (Nec) $\vdash F \Longrightarrow \vdash [O]F$ (if F is propositional and O a total order on a subset of $\{1,2\}$)
- (MP) $\vdash F$ and $\vdash (F \rightarrow G) \Longrightarrow \vdash G$

Notice that axioms $(A3)$ and $(A4)$ express the suspicious attitude of merging. Let us assume that the order is: $i_1 \geq i_2$. Then, in case of contradiction, the formulae which are true in the merged database are those which are true in database i_1. Else, the formulae which are true in the merged database are those of database i_1 or database i_2.

Remark: This logic will be used to model the merging of two databases $db1$ and $db2$, in the case where they are finite consistent sets of L literals.

Let us note

$$\psi = \bigwedge_{l \in bd1} [1]l \wedge \bigwedge_{bd1 \nvdash c} \neg[1]c \wedge \bigwedge_{l \in bd2} [2]l \wedge \bigwedge_{bd2 \nvdash c} \neg[2]c$$

(where l is a literal and c a clause)

We are interested in finding valid formulae of the form: $(\psi \rightarrow [O]F)$, i.e., formulae F which are true in the database obtained by merging $db1$ and $db2$, when the order is O.

We can prove the following facts:

Proposition 3. *Let $bd1$ and $bd2$ two finite sets of L literals and let ψ be the formula previously defined. Let F be a formula of L. Let O be a total order on a subset of $\{1,2\}$.*

$$FUSION - S(1,2) \models (\psi \rightarrow [O]F) \iff FUSION - S(1,2) \vdash (\psi \rightarrow [O]F)$$

This shows that deriving theorems of the form $(\psi \rightarrow [O]F)$ yields all the valid formulae of the form $(\psi \rightarrow [O]F)$.

Proposition 4. *Let $bd1$ and $bd2$ two finite sets of L literals and let ψ be the formula previously defined. Let F be a propositional formula. Let O be a total order on a subset of $\{1,2\}$. Then:*
$$FUSION - S(1,2) \vdash (\psi \rightarrow [O]F) \quad or \quad FUSION - S(1,2) \vdash (\psi \rightarrow \neg[O]F)$$

This shows a kind of completeness of the ψ consequences of the form $[O]F$: whether we will be able to derive that $[O]F$ (i.e., it is the case that F is true under order O), or $\neg[O]F$ (i.e., it is not the case that F is true under order O).

4.3 An example

Let us illustrate the suspicious attitude on the inspector example. Consider two witnesses. The first one, Bill, told the inspector that he saw a black car. The second one, John, told the inspector that he saw two men in a white car. So we consider the two following bases:

$1 = \{black\}$. $2 = \{two, \neg\ black\}$.

Then we consider $\psi = [1]\ black\ \wedge\neg[1]\ two\ \wedge\neg[1]\neg\ two\ \wedge\neg[1]\neg\ black\ \wedge[2]\ two$ $\wedge[2]\neg\ black\ \wedge\neg[2]\neg\ two\ \wedge\neg[2]\ black$

Here are some deductions that the inspector can perform:

$FUSION - S(1,2) \vdash (\psi \rightarrow [12]\ (black))$

$FUSION - S(1,2) \vdash (\psi \rightarrow \neg[12]\ (\neg\ black))$

$FUSION - S(1,2) \vdash (\psi \rightarrow \neg[12]\ two)$

$FUSION - S(1,2) \vdash (\psi \rightarrow \neg[12]\ (\neg\ two))$

$FUSION - S(1,2) \vdash (\psi \rightarrow [21]\ (two\ \wedge\neg\ black))$

$FUSION - S(1,2) \vdash (\psi \rightarrow \neg[21]\ black)$

In other terms, if the inspector trusts Bill more than John, the only thing he can assume is that the car was black. If he trusts John more than Bill, he can only assume that there were two men in a black car.

5 Merging Several Databases According to the Trusting Attitude

We still assume that the language L is a finite set of propositional variables.

We denote with $1, 2 \ldots n$ the n databases to be merged. Again, we assume that the databases are finite, satisfiable but not necessarily complete sets of literals.

The logic we define here is called $FUSION - T(1 \ldots n)$. Its pseudo-modalities are :

- $[i_1 i_2 \ldots i_m]$, where $m \geq 1$ and $i_j \in \{1 \ldots n\}$ and $(j \neq k \implies i_j \neq i_k)$

- $[i_1 i_2 \ldots i_m]\ F$ will mean that, when considering the total order on $\{i_1 \ldots i_m\}$: $i_1 > i_2, i_2 > i_3, \ldots i_{m-1} > i_m$, F is true in the database obtained by virtually merging database i_1 and $\ldots$ database i_m.

Notice again that the general form of these pseudo-modalities allows us to represent the particular case: $[i]F, i = 1 \ldots n$, which means that F is true in database i.

5.1 Semantics

The semantics of $FUSION - T(1 \ldots n)$ is the following: the model of $FUSION - T(1 \ldots n)$ is the pair: $I = (W, r)$, where:

- W is the finite set of all the interpretations of the underlying propositional language L

– r is a finite set of equivalence relations between interpretations in W.

Each pseudo-modality is associated with a relation. So, if $[O]$ is a pseudo-modality, $R(O)$ denotes the associated equivalence relation and $\overline{R}(O)$ its equivalence class. The equivalence classes are recursively defined by:

$\overline{R}(i)$ is a nonempty subset of $W, i = 1 \ldots n$

$\overline{R}(i_1 i_2 \ldots i_m)) = f_{i_m}(\ldots(f_{i_2}(\overline{R}(i_1)))\ldots))$ where:

$f_{i_j}(E) = \{w : w \in E \text{ and } w \models L_{i_j E}\}$ and
$L_{i_j E} = \{l : l \text{ literal of } L \text{ such that } : (\forall v \in \overline{R}(i_j) \Longrightarrow v \models l) \text{ and}$
$\qquad\qquad\qquad (\exists u \in E \text{ and } u \models l) \}$

Definition 5. Satisfaction of formulae.
(see Section 4.1)

Notice that since the underlying databases are assumed to be satisfiable, we have $\forall i = 1 \ldots n, \overline{R}(i) \neq \emptyset$, So, $\overline{R}(O) \neq \emptyset$, whatever the total order O on any subset of $\{1 \ldots n\}$. This means that (virtually) merging several satisfiable databases leads to a satisfiable database. In other terms, in our logic, not all propositional formulae will be true under an order O.

5.2 Axiomatics

Let O be an order $i_1 \ldots i_m, m \geq 1$. (i.e., $i_1 > i_2, \ldots, i_{m-1} > i_m$). By convention, we note $O \cup \{i_{m+1}\}$, the order $i_1 \ldots i_m i_{m+1}$.

Axioms of $FUSION - T(1 \ldots n)$ are:

- $(A0)$ Axioms of the propositional logic
- $(A1)$ $[O]\neg F \rightarrow \neg[O]F$
- $(A2)$ $[O]F \wedge [O](F \rightarrow G) \rightarrow [O]G$
- $(A3)$ $[O \cup \{i\}]l \longleftrightarrow [O]l \vee ([i]l \wedge \neg[O]\neg l)$ if l is a literal

Inferences rules of $FUSION - T(1 \ldots n)$ are:

- (Nec) $\vdash F \Longrightarrow \vdash [O]F$ (if F is a propositional formula)
- (MP) $\vdash F$ and $\vdash (F \rightarrow G) \Longrightarrow \vdash G$

Axiom $(A3)$ expresses the trusting attitude. Indeed, we could decompose it in three axioms:

- $(A3.1)$ $[O]l \rightarrow [O \cup \{i\}]\, l$
- $(A3.2)$ $[i]l \wedge \neg[O]\neg l \rightarrow [O \cup \{i\}]\, l$
- $(A3.3)$ $\neg[O]l \wedge \neg[i]l \rightarrow \neg[O \cup \{i\}]\, l$

This means that:

- if a literal is true under order O, then it is still true under order $O \cup \{i\}$, for any i, since, by convention, $O \cup \{i\}$ means that i is the least reliable source.

- if it is the case that a literal is true in database i, and if it is not the case that its negation is true under order O, then it is the case that it is true under $O \cup \{i\}$.

- if it is not the case that a literal is true under order 0 and if it is not the case that it is true in database i then it is not the case that it is true under order $O \cup \{i\}$.

Remark: This logic will be used to model the merging of n databases $db1 \ldots dbn$, in the case where such databases are finite consistent sets of L literals and according the trusting attitude

Let us note

$$\psi = \bigwedge_{i=1}^{n} (\bigwedge_{l \in bdi} [i]l \wedge \bigwedge_{bdi \nvDash c} \neg[i]c)$$

(where l is a literal and c a clause)

We will be interested in finding valid formulae of the form: $(\psi \rightarrow [O]F)$, i.e., formulae F which are true in the database obtained by merging $db1 \ldots dbn$, when the order is O.

We can prove the following facts:

Proposition 6. *Let $bd1 \ldots bdn$ be finite sets of literals and let ψ be the formula previously defined. Let F be a formula of L. Let O be a total order on a subset of $\{1 \ldots n\}$. Then:*

$$FUSION - T(1 \ldots n) \models (\psi \rightarrow [O]F) \iff FUSION - T(1 \ldots n) \vdash (\psi \rightarrow [O]F)$$

Proposition 7. *With the same assumptions as in proposition 3:*

$$FUSION - T(1,2) \vdash (\psi \rightarrow [O]F) \text{ or } FUSION - T(1,2) \vdash (\psi \rightarrow \neg[O]F)$$

5.3 An example:

Let us illustrate this machinery on the inspector example. Assume here that the inspector has got some belief about the context of the crime: in fact, the day of the crime was a foggy day. The inspector gathers information from two witnesses: Bill who told him he saw a black car and John who told him he saw two men in a white car. Let us consider three bases :

$1 = \{fog\}$. $2 = \{black\}$. $3 = \{two \wedge \neg black\}$.

We note $\psi = [1] fog \wedge \neg[1] black \wedge \neg[1] \neg black \wedge \neg[1] two \wedge \neg[1] \neg two \wedge \ldots [3] two \wedge [3] \neg black \wedge \neg[3] fog \wedge \ldots$

Here are some formulae that the inspector can deduce according to the trusting attitude:

$FUSION - T(1,2,3) \vdash (\psi \rightarrow [123] \ (fog \wedge black \wedge two))$

i.e., when assuming $1 > 2$, $2 > 3$, we can deduce that there were two men in a black car and there was fog.

$FUSION - T(1,3,2) \vdash (\psi \rightarrow [132] \ (fog \wedge two \wedge \neg \ black))$

i.e., when assuming $1 > 3$, $3 > 2$, we can deduce that there were two men in a car, that was black, and there was fog.

6 Links Between the Two Attitudes

Proposition 8. *Case of two databases.*[1] *When the two databases to be merged are compatible, then:*
$$FUSION - S(1,2) \vdash (\psi \rightarrow [O]F) \iff FUSION - T(1,2) \vdash (\psi \rightarrow [O]F)$$

i.e., when the two databases to be merged are compatible, the merged database obtained with the suspicious attitude is the same than the database obtained with the trusting attitude.

7 Relation with the Problems of Database Updates and Belief Base Updates

Let us first recall some definitions (Katsuno and Mendelzon 1991):

Definition 9. Let m and $m1$ be two interpretations. We define the distance between m and $m1$ by:
$$d(m, m1) = \{p \in L : (p \in m \text{ and } p \notin m1) \text{ or } (p \notin m \text{ and } p \in m1)\}$$

Definition 10. Let m be an interpretation. We define a partial order on the set of interpretations $\leq_m$ by:
$$m1 \leq_m m2 \iff d(m, m1) \subseteq d(m, m2)$$

Then we have the following result:

Proposition 11. *Let f_{i_j} be the functions defined in Section 5.1. Then:*
$$f_{i_j}(E) = \bigcup_{m \in \overline{R}(i_j)} Min(E, \leq_m)$$

So, according to (Katsuno and Mendelzon 1991), $f_{i_j}(E)$ is the set of models of belief base i_j updated by E.

In other terms, the result of merging two databases according to the trusting attitude, such that $i > j$, is equivalent to updating database j with database i.

This also means that the axiomatics given for two sets of literals in Section 5.2 is an axiomatics of an update of atomic belief bases i.e.,
$\{F : FUSION - T(1,2) \vdash [ij]F\} \longleftrightarrow j <> i$ (notation of Katsuno and Mendelzon (1991)).

---

[1] With the previous notations.

Generalisation: Merging n databases $1, 2 \ldots n$, according to the trusting attitude and given an order: $i_1 > i_2, \ldots, i_{n-1} > i_n$ means to update i_n with the result of the update of i_{n-1} with the result of the update ... of i_2 with i_1, i.e.,
$$\overline{R}(i_1 \ldots i_{n-1} i_n) = \text{models-of}(i_n <> (i_{n-1} <> \ldots (i_2 <> i_1) \ldots))$$

Reconsider Winslett's approach (Winslett 1990). The previous proposition shows that adding a conjunction i in a database j according to Winslett's approach, means the merging of i and j, according to the trusting attitude in the order $i > j$.

Besides that, in the classical approaches to database updates, the addition of any information i that contradicts the current state of the database j is rejected. This implies the merging of i and j according to the suspicious attitude and in the order $j > i$.

8 Concluding Remarks and Open Problems

A query evaluator that implements the logics presented here has been developped (Cholvy 1993). It considers queries of the form: $[O]Q$? i.e., "is Q is true under order O?".
It generates three kinds of answers:
  * YES, if $FUSION - S(1, 2) \models [O]Q(resp, FUSION - T(1, n))$
  * NO, if $FUSION - S(1, 2) \models \neg[O]Q(resp, FUSION - T(1, n))$
  * ?, else

However, the work presented here leaves many problems open:

First, the logics presented here must be extended in the case where databases are sets of clauses. In this context, given the semantics of the trusting attitude, we will be constrained to consider the second definition of $f_{i_j}(E)$ in terms of minimal models (see Section 7) since the first definition given in Section 5.1 can not be applied generally.

Second, we have shown that merging several databases may be achieved according to two attitudes. The so-called trusting attitude corresponds to consecutive updates. So postulates of updates (when restricted to our case of conjunctions) characterize merging. It would be interesting to see what are the postulates that characterize the suspicious attitude.

Finally, we have assumed here that the databases are ordered according to a total order. However, in some applications, the person who gathers the different sources of information cannot order them totally: for instance, the inspector cannot decide which witness is more trustable. In the case of a partial order, we must define a formalism which mixes the logics presented here and the formalisms defined for reasoning in inconsistent contexts with equally trusted information (see Section 2).

References

Besnard P.: "Logics for automated reasoning in the presence of contradictions," in: *Proceedings of Artificial Intelligence: Methodology, systems and applications.* North Holland, 1990.

Bauval A., and L. Cholvy: "Automated reasoning in case of inconsistency," in: *Proceedings of the First World Conference on Fundamentals of AI* 1991.

Baral C., S. Kraus, J. Minker, and V.S. Subrahmanian: "Combining knowledge bases consisting of first order theories," in: *Computational Intelligence* **8-1** (1992).

Cayrol C., R. Royer, and C. Saurel: "Management of preferences in assumption-based reasoning," in: *Proceedings of (IPMU) Conference.* Palma de Mallorca, Spain, 1992.

Cholvy L.: "Querying an inconsistent database," in: *Proceedings of Artificial Intelligence: Methodology, systems and applications.* North Holland, 1990.

Cholvy, L.: "Proving theorems in a multi-source environment," in: *Proceedings of IJCAI*, 1993.

Dubois D., J. Lang, and H. Prades: "Dealing with multi-source information in possibilistic logic," in: *Proceedings of ECAI*, 1992.

Fagin R., J. Ullman, and M. Vardi: "On the semantics of updates in databases," in: *Proceedings of ACM TODS*, 1983.

Fagin R., G. Kupper, J. Ullman, and M. Vardi: "Updating logical databases," in: *Advances in Computing Research* **3** (1986).

Farinas Del Cerro L., and A. Herzig: *Reasoning about database updates. Proceedings of the Workshop of Foundations of Deductive Databases and Logic Programming.* Jack Minker Editor, 1986.

Farinas L., and A. Herzig: *Constructive Minimal Changes.* Report IRIT, 1992.

Gabbay D., and A. Hunter: "Making inconsistency respectable," in: *International Workshop on Fundamentals of Artificial Intelligence*, 1991.

Katsuno H., and A. Mendelzon: "Propositional knowledge base revision and minimal change," in: *Artificial Intelligence* **52** (1991).

Winslett M.: *Updating Logical Databases.* Cambridge University Press, 1990.

Situation Theory and Social Structure

Keith Devlin

School of Science, Saint Mary's College of California
Moraga, California 94575, USA
E-mail: devlin@stmarys-ca.edu

Abstract. This paper seeks to utilize situation theory as a framework
(or formalism) for the description and analysis of the fundamental social
structures that underlie the way we encounter the world, and which
influence our behavior and communication in society.

Much of this work is done jointly with Duska Rosenberg. Our collaboration on this project commenced with (Devlin and Rosenberg, to appear),
and continues in (Devlin and Rosenberg, in preparation a). In many
ways, this paper represents a partial progress report of work leading to
our forthcoming monograph (Devlin and Rosenberg, in preparation b).

1 Introduction

The traditional approach to social study, sometimes referred to as 'normative
sociology', commences with a collection of empirically identified *social norms*
that constitute, or are taken to constitute, our common-sense-view of the world.
Normative sociology posits, and then attempts to describe, an objective world of
social facts to which our attitudes and actions are a response. What the everyday
practice of normative sociology amounts to is constructing a social science by a
process of refinement, or quantification, of the initial collection of foundational
norms. This enterprise is successful in so far as it improves upon the normative
structure it starts with. Human action is explained by reference to the body of
social norms taken to be the grounding structure for the theory.

During the 1960s, Garfinkel (1967) and others proposed a radically different
approach, known as *ethnomethodology*. Rather than taking a normative social
structure as foundational and explain human action in terms of those norms, the
ethnomethodologist regards human action as fundamental and seeks to explain
how human action can give rise to what we perceive as a collection of norms.
Thus for the ethnomethodologist, the common-sense-view of the world is taken
not as a foundational structure to be improved upon, but as the fundamental
phenomenon of social study.

The ethnomethodologist, then, seeks to identify, describe, and understand the
social *structure* that underlies the way ordinary people (as opposed to experts
or theorists) encounter the world and make it intelligible to themselves and to
one another. As such, ethnomethodology seeks to provide:

descriptions of a society that its members, (...) use and treat as known in common
with other members, and with other members take for granted. Specifically, [it seeks]

a description of the way decisions of meanings and fact are managed, and how a body of factual knowledge of social structures is assembled in common sense situations of choice. (Garfinkel 1967: 77.)

By way of an example, consider an utterance of the following sentences:

The baby cried. The mommy picked it up.

These two sentences, uttered by a small child as the opening two sentences of a story, constitute the principal data of a well-known, seminal article in the ethnomethodological literature, Harvey Sacks' *On the Analyzability of Stories by Children* (1972), an article which Rosenberg and I subjected to a situation-theoretic analysis in (Devlin and Rosenberg, to appear).

In his article, Sacks is concerned with natural conversation, and in particular with the way speaker and listener make use of their knowledge of social structure in the utterance and understanding of a simple natural language communication. According to Sacks, the particular choice of words used by a speaker in, say, a description is critically influenced by the speaker's knowledge of social structure, and the listener utilizes his[1] knowledge of social structure in order to interpret, in the manner the speaker intended, the juxtaposition of these words in conversation.

The focus of Sacks' article is the way a *typical* listener understands the two sentences. As Sacks observes, when heard by a typical, competent speaker of English, the utterance is almost certainly heard as referring to a very small human (though the word 'baby' has other meanings in everyday speech) and to that baby's mommy (though there is no genitive in the second sentence, and it is certainly consistent for the mommy to be some other child's mother); moreover it is the baby that the mother picks up (though the 'it' in the second sentence could refer to some object other than the baby). Why do we almost certainly, and without seeming to give the matter any thought, choose this particular interpretation?

To continue, we are also likely to regard the second sentence as describing an action (the mommy picking up the baby) that follows, and is caused by, the action described by the first sentence (the baby crying), though there is no general rule to the effect the sentence order corresponds to temporal order or causality of events (though it often does so).

Moreover, we may form this interpretation without knowing what baby or what mommy is being talked of.

Furthermore, we recognize these two sentences as constituting a *possible description*, and indeed it seems to be in large part because we make such recognition that we understand the two sentences the way we do.

As Sacks notes, what leads us, effortlessly, instantaneously, and almost invariably, to the interpretation we give to this simple discourse, is the speaker and listener's shared knowledge of, and experience with, the social structure that pertains to (the subject matter of) this particular utterance.

---

[1] For definiteness, I shall assume a female speaker and a male listener throughout this paper.

It is this underlying social structure that constitutes Sacks' main interest. Referring to his observations concerning the way typical speakers of English understand the two sentences, summarized above, Sacks says of the social structure he (and we) seek to investigate (1972: 332):

My reason for having gone through the observations I have so far made was to give you some sense, right off, of the fine power of a culture. It does not, so to speak, merely fill brains in roughly the same way, it fills them so that they are alike in fine detail. The sentences we are considering are after all rather minor, and yet all of you, or many of you, hear just what I said you heard, and many of us are quite unacquainted with each other. I am, then, dealing with something real and something finely powerful.

Having identified a phenomenon to be studied, as Garfinkel, Sacks, and the other early ethnomethodologists surely did, the question at once arises as to how one should set about studying that phenomenon. Ethnomethodologists insist that the essence of the discipline is *inquiry*, not theorizing; the ethnomethodologist examines the (empirical) data, holding them up for scrutiny and analysis; she does not use the data as a basis for theory-building. (See, for instance, the remarks of Button (1991: 4–5).)

But how is that examination, that analysis of the data, to be carried out, and how are the ethnomethodologist's findings to be evaluated and validated? Clearly, both description and analysis must, of necessity, be carried out with the aid of language. But then we run into the problem that will inevitably face any deeply foundational study of linguistic issues, what one might term the 'epistemological-bootstrap' problem: successfully using language in order to examine, and communicate to others the results of such examination, the *minutae* by which language functions as a communicative medium.

One need only look at, say, the Sacks' paper to see the difficulties such a fundamental study entails. The ethnomethodological literature is riddled with long, highly complex sentences, often extremely hard to parse, and authors frequently resort to the construction of complex, hyphenated terms in order to try to capture adequately the relevant abstract concepts they wish to examine. Though perhaps frustrating for the reader of an ethnomethodological analysis, this is, in large part, surely unavoidable. As Benson and Hughes argue in (Button 1991: 129):

... sociology is, along with the other human sciences, a 'second order' discipline using the concepts of ordinary language. In which case, an important constraint on sociological inquiry is that the categories, the concepts used, and the methods for using them, must be isomorphic to the ways in which they are used in common-sense reasoning.

The absence of a descriptive and analytic framework 'separate' from the data under consideration, while perhaps a source of frustration to a sociologist trying to follow an ethnomethodological analysis, can present a major obstacle to interdisciplinary research, where people with different areas of expertise wish to pool their knowledge to attack particularly difficult problems; for example, the kind of interdisciplinary research and development work involved in the design of interactive information systems in the context of Computer

Supported Cooperative Work (CSCW), a process in which ethnomethodology is playing an increasingly significant role these days. The relative ease with which say, the chemist, biologist, and physicist can communicate with each other at a fundamental level through the universal language of mathematics and the use of a common methodology (the 'scientific method'), is not matched when an ethnomethodologist becomes part of a systems-design team involving computer scientists, mathematicians, and engineers.

Can some measure of assistance be found through increased 'formality' of language? More specifically, can mathematical formalisms be advantageously introduced?

It is the purpose of this paper, just as it was the purpose of my joint paper with Rosenberg (Devlin and Rosenberg, to appear), to argue that mathematics can indeed be used profitably in carrying out an analysis of phenomena normally alalyzed by ethnomethodologists. That this claim may strike many as *quite obviously false* is, I would suggest, a consequence of a mistakenly narrow view of what mathematics is, or at least what it can be and what it can be used for,[2] an issue I take up in the next section.

The crucial factor to be addressed in attempting to make use of mathematical techniques in this kind of analysis is to fit the mathematics to the data, not the other way round. And that almost certainly requires the development of new mathematical techniques.

My paper with Rosenberg (Devlin and Rosenberg, to appear) presents an analysis of Sacks' article. We attempt to show how situation theory can help formalize some of Sacks' arguments, and we make use of that formalization to examine those arguments in a critical fashion. Our main purpose was not to present an alternative analysis of the same linguistic data, but to demonstrate that situation theory could be used to add precision to an ethnomethodological analysis *already validated as such within the ethnomethodological community.* The added precision we were able to bring to Sacks' arguments brought out some points of his argument that were in need of clarification, a clarification that we were, again by virtue of the situation-theoretic machinery, able to provide. In the course of this study, we were led to new structures and new classes of situation-theoretic constraints, not previously considered in situation theory.

In this present paper, I shall give a brief summary of the discussion of the Sacks' article Rosenberg and I presented in (Devlin and Rosenberg, to appear), taking the situation-theoretic development we presented there a stage further, in particular, showing how situation theory can help describe the way that the constraints that govern normal behavior can arise through interactions in a social context. This leads to the introduction of a new category of situation-theoretic *types* not previously encountered in situation theory. Using these types—I refer to them as *evolved types*—I sketch an alternative (and in my view superior)

---

[2] A view which, I am afraid, is due in part to the manner in which my fellow mathematicians have traditionally regarded both their subject and its relation to other disciplines, particularly the humanities.

situation-theoretic analysis of the Sacks data than the one Rosenberg and I presented in (Devlin and Rosenberg, to appear).

The machinery I develop here sets the stage for a situation-theoretic investigation of the issue of *possible descriptions*: what is it about certain sequences of utterances that enables agents such as ourselves to recognize, without effort, a sequence of utterances as constituting, or possibly constituting, a description of some object, scene, or event? Though a discussion of this notion occupies a good deal of Sacks' paper, Rosenberg and I did not consider this aspect of Sacks' analysis in (Devlin and Rosenberg, to appear); rather we left that as the topic to be taken up in a later paper (Devlin and Rosenberg, in preparation a), which we are currently working on.

To skeptics that are tempted to read no further, let me at this stage simply quote one of the remarks Sacks makes early in his *mother–baby* article, a remark also addressed to anticipated skeptical readers (Sacks 1972: 330):

... I ask you to read on just far enough to see whether it is or is not the case that the observations [I make] are both relevant and defensible.

2 Mathematics and Sociology

The article by Benson and Hughes (Button 1991: Chapter 6) summarizes the most well-known application of mathematical techniques in sociology, the 'variable analysis' commonly associated with Lazarsfeld. In this approach to social study, various features of a society are denoted by numerical parameters, whose values are then obtained by empirical, investigative techniques (tabulating and counting). As a *statistical* approach, variable analysis can provide useful insights into the behavior of *sections* of a society, consisting of many members, and it may well be possible, on occasion, to particularize from such global findings to understand, and perhaps even predict, aspects of behavior of individual members of such a section. But, being quantitative, it is difficult to see how such techniques could be of much use to ethnomethodology, which concentrates on fundamental, cognitive and social behavior at the level of individuals.

Given the extensive use made of statistical techniques in sociology—in 1971, one commentator estimated (Phillips 1971) that some 90% of published empirical sociology was produced using the statistical manipulation of interview and questionnaire data—it is not surprising that to many sociologists, the phrase 'mathematical techniques' is virtually synonymous with 'statistical (or quantitative) techniques'. For instance, in the article mentioned above, Benson and Hughes claim that mathematical techniques are not appropriate in ethnomethodology, but their argument applies only to *quantitative* techniques, which they appear to confuse with the far more general notion of a 'mathematical technique'. Thus, on page 117 of (Button 1991), we read:

To the extent to which a mathematical system is applied, 'objects' in the target domain must be mapped onto 'objects', *that is numbers*, in the mathematical domain. [My emphasis.]

The authors seem to make two major assumptions here. Firstly, they (wrongly) assume that the only mathematical objects that might be used are numbers. And, even more fundamental, they apparently believe (again erroneously) that a mathematical system can only deal with abstract mathematical objects (be they numbers or other), and not with the same, real objects, agents, and data that the ethnomethodologist is mostly concerned with in her everyday work.

Clearly, then, before we can take another step, we need to be clear what is meant (certainly what I mean) by the word 'mathematics'.

Present-day mathematics evolved in eras when quantization was paramount: commerce, architecture, cartography, engineering, and the laboratory sciences all required numerically-based tools for measurement and computation. In contrast, the kind of mathematics I would suggest might be of assistance in an ethnomethodological analysis would be quite different, based on descriptive rather than numerical concepts. The relevant issues are essentially *descriptive* and *qualitative*, and are, as Benson and Hughes observe, not amenable to techniques that are ultimately *quantitative*. What is required then is *qualitative* or *descriptive* mathematics, not a *quantitative* theory.[3]

Certainly, the *development* of various kinds of mathematics to meet a particular need is not a new phenomenon. The evolution of mathematics is a long series of developments made in response to needs of the time: arithmetic to facilitate trade and commerce among the Sumerians, the Babylonians, and the Egyptians, geometry and trigonometry to support exploration and the construction of buildings by the Greeks, the calculus to allow for a precise study of motion and space in physics in the 17th Century, projective geometry to understand how a three-dimensional world may be represented on a two-dimensional canvas during the Renaissance period, techniques in calculus to meet the demands of structural and (later) electrical engineers in the 19th and early 20th Centuries, parts of logic and discrete methods to facilitate the development of computers in the 20th Century, and more recently developments in topology and knot theory to help understand the genetic material DNA.

There is, as far as I can see, no *a priori* reason that precludes the development of mathematical tools for use in sociology, including ethnomethodology. How much these tools resemble existing parts of mathematics is not clear, and is quite unimportant. What is important is that the machinery will have to be developed with the intended application firmly in mind, by carrying out the kind of study presented in this paper. In particular, the mathematical tools we produce must be designed and used in such a way that their formality does not constrict the application domain. This involves a careful choice of both the *level* at which the

---

[3] Though mathematics has always had qualitative aspects, the quantitative aspects have hitherto dominated, particularly in applications. Indeed, so fundamental has been number and measurement to most of mathematics over its 5,000 year history, that a great deal of the descriptive mathematics that has been used in, say, computer science and linguistics has had to be developed almost from scratch, with those applications in mind, largely based on mathematical logic, one of the few purely qualitative branches of classical mathematics.

mathematics is developed and applied, and the *granularity* of the mathematical framework that is developed at that level. (For instance, Euclidean geometry is an excellent mathematical tool for studying the geometry of the everyday world around us, but not at all suitable for studying the universe on either a cosmic or a subatomic level, when other descriptive mechanisms have to be used.)

The development of mathematical tools to be used in any discipline proceeds by a process of abstraction: fleshing out the abstract structures that underlie the phenomena under consideration, in a fashion consistent with the aims and overall methodology of that discipline. In the case of mathematical tools for use in ethnomethodology, attention needs to be paid to the intrinsic methodology of that particular approach to the study of social phenomena. The aim must be to abstract a mathematical framework *endogenous* to the system, not to port or adapt some mathematical framework developed for other purposes, least of all a framework developed for applications in the natural sciences.

Of course, the result of such a development may well be a mathematical theory that does not look very much like most existing branches of mathematics. If this is the case, then so be it. As Lynch says in (Button 1991: 97):

The policy of ethnomethodological indifference [to 'method'] requires that we put aside strong professional urgings to 'ground' inquiry in a set of *a priori* rules and standard techniques ...

3 Situation Theory

Though developed initially by a logician (Jon Barwise), situation theory is not, as some have supposed, an attempt to extend classical logic to cover various real-world phenomena, in particular the use of natural language in communication. Such a *bottom-up*, mathematical-logic treatment of natural language has indeed been attempted, most notably by Chomsky, in the case of grammar and syntax, and by Montague in the case of semantics. In contrast, situation theory adopts a very definite *top-down* approach. Starting from the assumption that purposeful action and communication require a certain 'orderliness' in the world, situation theory attempts to develop—or more accurately *abstract*—a conceptual framework with which to describe and study that orderliness.

As a top-down development, situation theory deals with real objects, real agents, and real data, identifying and gradually refining the various abstract structures that arise from, and govern, the behavior and actions of members of a society.[4] (In contrast, bottom-up approaches generally start out with a collection of abstract, mathematical objects, often pure sets and functions, and gradually add structure until a model of the target phenomenon is obtained. In such an approach, the abstract mathematical objects are eventually taken to *represent*, in some sense, various objects and entities in the world.)

---

[4] Hitherto, the 'societies' considered have been linguistic communities or subcommunities thereof, including the case of computers and interactive information-processing systems.

In starting with empirical data, and gradually abstracting an appropriate (for our purposes) framework, we take care, at each stage, to make only the minimal restrictions and refinements necessary to proceed. Though formalization (in the usual sense of this word in mathematics) and axiomatization might well be ultimate goals (at least for some), the important issue is to remain as true to the data as possible; and every new restriction, every new tightening up of a definition, runs the risk of distorting the data or even excluding some features from further consideration. Such a cautious, painstaking, and in some ways deliberately non-committal approach is intensely frustrating to most mathematicians, including myself, but is absolutely necessary if there is to be any hope of achieving the desired aim.

A similar point is made by Jeff Coulter (Button 1991: 39–40), in connection with what he calls 'endogenous logic':[5]

The relative neglect of *symbolic* formalization and axiomatization within Endogenous Logic will not be discussed here: suffice to say, much of the narrowness of scope of traditional formal–logical studies of language may be attributed to attempts to preserve the consistency of a notation system at the expense of discerning fresh and actual logical relationships and connections orientable-to, and made by, practical reasoners in the course of conducting their everyday affairs. Further, an exclusive reliance upon supplementing the classical logical (particularly 'formal semantic') concepts with mathematical ones (especially the one of 'set'), and a related (Carnapian) insistence upon distinguishing between a 'logical' and a 'nonlogical' vocabulary or conceptual apparatus in natural languages, can both be seen now as unnecessary restrictions upon the logical investigation of language-use and cultural *praxis*. The exact extent of the revisions necessary in relation to former conceptions of logical analysis remains an open question.

Of course, even if we decide (as I do) to study, in a top-down fashion, the everyday world of objects, agents, spatial and temporal locations, and the like, that ordinary people see and experience as making up their *environment*, there has to be some initial set of assumptions concerning the kinds of abstract structures to be considered. Something has to distinguish our study from physics, chemistry, biology, psychology, and even from normative sociology and Garfinkel-style ethnomethodology. Though we are, as a driving methodology, taking the world (our data) as it is cut up by the ethnomethodologists, we are, after all, attempting to develop a new way of *looking at* that data. Situation theory is intended to provide a useful alternative view of issues such as linguistic communication and human interaction.

Situation theory is a conceptual framework for the study of *information*. When used as a framework for the study of social phenomena, the result is an

---

[5] There is as yet no such subject as 'endogenous logic'. Rather Coulter has proposed the development of such a logic, and has set out some guidelines for that development. The relationship between situation theory and Coulter's *desiderata* for such a logic is considered in Section 8.

information-theoretic approach to communication and action. To explain what these two statements mean, I refer to Fig. 1.

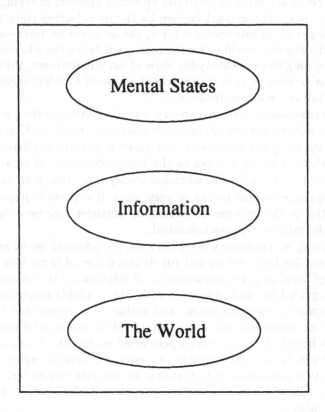

Fig. 1. The information level

In order to study cognition, communication, and action *in a manner that accepts our everyday view of the physical world*, we need to take account of two distinct domains: the (physical) world on the one hand, and mental states on the other.[6] In the case of the former, the physical world, mathematics can provide us with a number of proven techniques and well-worked-out models. Indeed, this is the domain in which the development and use of mathematical techniques has been supremely successful. The same cannot be said for the other domain, the domain of mental states. Though research in artificial intelligence and in cognitive science has attempted to develop the appropriate machinery, this effort has, to date, been largely unsuccessful. Indeed, I see no reason to

---

[6] I am making some element of judgment here. Behaviorism could be said to be an attempt to circumvent the requirement of a model of mental states, by reducing all cognitive activity to the externally manifested stimulus–response activity of the agent. However, few workers these days believe that this approach would be adequate for the kind of purposes outlined in this paper.

assume that mathematical techniques will ever be as successful in providing models of cognitive activity as they have been in modeling the physical world.

Situation theory attempts to avoid the problems involved in trying to model mental activity in a mathematical fashion by the introduction (into the mathematical ontology) of an intermediate layer, the information layer—see Fig. 1. Indeed, since the physical world is, for the most part, taken (by situation theory) to be the one arising from our everyday view of our environment, that is to say, the world of naive physics, it is really only at the level of information that the theory is an abstract, *mathematical* one.

And at the information level, our theory is mathematical to the core. Though the intuitions concerning information that motivate the mathematical treatment are rooted in our everyday experience, our entire apparatus for handling information is mathematical, right down to the formal definition of what we mean by 'information' (or at least, what situation theory takes information to be).

In this respect, situation theory is very much like particle physics, which takes the matter of the universe to consist of 'particles', whose only real and unchallengeable 'existence' is mathematical.

By postulating an ontological level between the physical world and mental states, we do not (at least, we try not to) distort either of those two aspects of reality. And by developing the mathematics of information in a manner that is motivated by, guided by, and consistent with, the available empirical evidence concerning cognition, communication, and action, it is reasonable to suppose that the resulting framework will be one that leads to meaningful results. More precisely, it is hoped that an analysis performed within the framework of situation theory will be one that reliably increases our understanding of various phenomena. (As a *mathematical* framework, an analysis carried out within situation theory will, of course, be 'universally accessible' and open to detailed inspection by others.)

This does not in any way make situation theory a 'theory of everything', not even everything in the cognitive domain. But it does, I believe, result in a framework that has considerable potential utility.

I am very aware that such an approach, indeed any approach that adopts a particular methodological 'stance', risks the fate that, according to Benson and Hughes (Button 1991: 125), befalls variable analysis:

... the burden of the ethnomethodological critique is, as far as variable analysis is concerned, that it sets the conditions for sociological description in advance so that we look at the phenomena through a grid that we impose upon them, *irrespective of whatever properties the phenomena might otherwise display.* [Emphasis as in the original.]

Stepping the path between adopting a particular viewpoint to gain understanding, and allowing the particulars of that viewpoint obscure what is being viewed, is a delicate one, and the best one can do is to keep referring back to the data as it is, and not how we might like it to be. This is, after all, what is done in the natural sciences. In physics, for example, there is no doubt that one is working not with 'the real, physical universe', but with a mathematical idealization, or model, of that domain. What makes the resulting theory so useful is that the physicist constantly compares her model with data from the real world. Occasionally, a discrepancy is found, and the model (not the world) has to be changed, as happened when Newtonian mechanics was found lacking and supplanted (for certain purposes) by the theories of relativity and quantum mechanics. In itself, there is nothing wrong with adopting a particular viewpoint or framework; indeed, without it, it is hard to see how any progress (in developing our understanding) could be made. This procedure only breaks down when the theoretical stance or framework continues to be used in circumstances discovered to be inappropriate.

To the critic who says that such a 'rationalist' (or Leibnizian) approach is not suitable in the case of ethnomethodology, I would say two things:

Firstly, use of a tool does not require a commitment to the framework within which that tool was developed. What counts is how effective that tool is in carrying out the task at hand. Just as the sociologist, psychologist, or phenomenologist, may make use of a computer in her analysis of human activity, without committing to the view that the computer (a Leibnizian device if ever there were one) provides a good model of human cognitive behavior, so too the argument Rosenberg and I present in (Devlin and Rosenberg, to appear) demonstrates, at the very least, that the ethnomethodologist may use situation theory as an effective, analytic tool, without such use requiring or entailing any commitment to our philosophy.

Secondly, it is surely foolish to turn one's back on *any* means to achieve greater understanding of significant aspects of the problem relevant to the purpose of the analysis. Faced with a highly complicated phenomenon—and in studying human behavior, the sociologist is surely faced with a phenomenon as complicated as it gets—the better ways one can find to look at that phenomenon, the greater will be the final understanding.

Presented with an unfamiliar object, the most sensible way to gain a proper understanding of that object is to view, examine, and analyze it in as many different ways as possible, using whatever tools we have at our disposal: look at it from all sides, pick it up, weigh it, X-ray it, take it apart, dissect it, analyze its chemical and physical composition, etc.

Likewise, a well-taken photograph of a country scene, being an accurate *visual* portrayal, will provide me with a pretty good sense of that scene. But my appreciation will be greater, and deeper, if I am also shown a good oil-painting, through which medium the artist can convey many of the aspects of the scene that a photograph cannot capture, aspects relating to mood and emotion and visual texture. A recording of the sounds at the location will provide still further understanding. And greater still will be my appreciation, if the picture and the painting and the sound-recording are accompanied by a poem or a piece of well-written prose, or even, on occasion, by a well-chosen (or inspirationally written) piece of music. (Notice that the painting, sound-recording, poem, prose, or music need not be acknowledged as good *within their respective disciplines* in order to be efficacious in contributing to my overall appreciation of the scene, which comes from a holistic synthesis of all the different portrayals.)

If the aim is to achieve as good an understanding as possible—and this, I take it, is the purpose of interdisciplinary research—then there is likely to be something to be gained by any additional means by which we can describe and analyze the phenomenon of interest. Situation theory is intended to provide one method by which one may gain a measure of understanding of communication and action.

In particular, situation theory offers us better understanding because it forces us to clarify issues that the social science perspective alone will (of necessity) take as given. Taken together, the two approaches make us examine and question issues that would otherwise not be noticed as problematic and non-trivial.

NOTE: *The remainder of this section consists of a brief summary of the basic ideas of situation theory. As such, the exposition is fairly technical, and the reader might prefer to skip ahead to the next section, and come back to this account as and when it proves necessary. A more leisurely, and comprehensive account of elementary situation theory is provided in (Devlin 1991.)*

Situation theory takes its name from the mathematical device introduced in order to take account of context and partiality. A *situation* can be thought of as a limited part of reality. Such parts may have spatio-temporal extent, or they may be more 'abstract,' such as fictional worlds, contexts of utterance, problem domains, mathematical structures, databases, or Unix directories. The status of situations in the ontology of situation theory is equal to that of (say) individuals; the distinction between the two being that situations have a *structure* that plays a significant role in the theory whereas individuals do not. Examples of situations of particular relevance to the subject matter of this paper will arise as the development proceeds.

The basic ontology of situation theory consists of entities that people, or other kinds of agent, individuate and/or discriminates as they make their way in the world: spatial locations, temporal locations, individuals, finitary relations, situations, types, and a number of other, 'higher-order' entities, together with some theorist's conveniences such as parameters.

The objects (known as *uniformities*) in this ontology include the following.

- *individuals* — objects such as tables, chairs, tetrahedra, cubes, people, hands, fingers, etc. that the agent either individuates or at least discriminates (by its behavior) as single, essentially unitary items; usually denoted in our theory by $a, b, c, \ldots$.
- *relations* — uniformities individuated or discriminated by the agent that hold of, or link together specific numbers of, certain other uniformities; denoted by $P, Q, R, \ldots$.
- spatial *locations*, denoted by $l, l', l'', l_0, l_1, l_2$, etc. These are not necessarily like the 'points' of mathematical spaces (though they may be so), but can have spatial extension.

– *temporal locations*, denoted by $t, t', t_0, \ldots$. As with spatial locations, temporal locations may be either points in time or regions of time.
– *situations* — structured parts of the world (concrete or abstract) discriminated by (or perhaps individuated by) the agent; denoted by $s, s', s'', s_0, \ldots$
– *types* — higher order uniformities discriminated (and possibly individuated) by the agent; denoted by $S, T, U, V, \ldots$.
– *parameters* — indeterminates that range over objects of the various types; denoted by $\dot{a}, \dot{s}, \dot{t}, \dot{l}$, etc.

The framework that 'picks out' the ontology is referred to as the *scheme of individuation*. The intuition is that in a study of the activity (both physical and cognitive) of a particular agent or species of agent, we notice that there are certain regularities or *uniformities* that the agent either individuates or else discriminates in its behavior.

For instance, people individuate certain parts of reality as *objects* ('individuals' in our theory), and their behavior can vary in a systematic way according to spatial location, time, and the nature of the immediate environment ('situation types' in our theory).

Information is always taken to be information *about* some situation, and is taken to be in the form of discrete items known as *infons*. These are of the form

$$\ll R, a_1, \ldots, a_n, 1 \gg , \; \ll R, a_1, \ldots, a_n, 0 \gg$$

where R is an n-place relation and $a_1, \ldots, a_n$ are objects appropriate for R (often including spatial and/or temporal locations). These may be thought of as the informational item expressing that objects $a_1, \ldots, a_n$ do, or respectively, do not, stand in the relation R.

Infons are 'items of information'. They are not things that in themselves are true or false. Rather a particular item of information may be true or false *about a certain part of the world* (a 'situation').

Infons may be combined to form what are known as *compound infons*: the permissible combinatory operations consist of conjunction, disjunction, and bounded universal and existential quantification (by parameters). I omit the details in this brief summary.

Given a situation, s, and an infon σ, we write

$$s \models \sigma$$

to indicate that the infon σ is 'made factual by' the situation s, or, to put it another way, that σ is an item of information that is true of s. The official name for this relation is that s *supports* σ. The facticity claim $s \models \sigma$ is referred to as a *proposition*.

It should be noted that this approach treats 'information' as a *commodity*. Moreover a commodity that does not have to be 'true.' Indeed, for every positive infon there is a dual negative infon that can be thought of as the 'opposite' informational item, and both of these cannot be 'true.'

The *types* of the theory are defined by applying two type-abstraction procedures, starting with an initial collection of *basic types*. The basic types correspond to the process of individuating or discriminating uniformities in the world at the most fundamental level. Among the *basic types* are:

TIM : the type of a temporal location;
LOC : the type of a spatial location;
IND : the type of an individual;
REL^n : the type of an *n*-place relation;
SIT : the type of a situation;
INF : the type of an infon;
PAR : the type of an *parameter*.

For each basic type T other than *PAR*, there is an infinite collection $T_1, T_2, T_3, \ldots$ of *basic parameters*, used to denote arbitrary objects of type T. We generally use the less formal notation $\dot{l}, \dot{t}, \dot{a}, \dot{s}$, etc. to denote parameters (in this case of type *LOC, TIM, IND, SIT*, respectively).

Given an object, x, and a type, T, we write

$$x : T$$

to indicate that the object x is *of* type T.

Most uses of parameters require what are known as *restricted parameters* whose range is more fine grained than the basic parameters. In essence, the mechanism for constructing restricted parameters enables us to make use of parameters restricted to range over any relevant domain. See (Devlin 1991) for details.

There are two kinds of type-abstraction, leading to two kinds of types. First of all there are the *situation-types*. Given a *SIT*-parameter, $\dot{s}$, and a compound infon σ, there is a corresponding *situation-type*

$$[\dot{s} \mid \dot{s} \models \sigma],$$

the *type* of situation in which σ obtains.

This process of obtaining a type from a parameter, $\dot{s}$, and a compound infon, σ, is known as *(situation-) type abstraction*. We refer to the parameter $\dot{s}$ as the *abstraction parameter* used in this type abstraction.

For example,

$$[SIT_1 \mid SIT_1 \models \ll \text{running}, \dot{p}, LOC_1, TIM_1, 1 \gg]$$

(where $\dot{p}$ is a parameter for a person) denotes the type of situation in which someone is running at some location and at some time. A situation s will be of this type just in case someone is running in that situation (at some location, at some time).

As well as situation-types, our theory also allows for *object-types*. These include the basic types *TIM, LOC, IND, REL^n, SIT*, and *INF*, as well as the more fine-grained uniformities described below.

Object-types are determined over some initial situation. Let s be a given situation. If $\dot{x}$ is a parameter and σ is some compound infon (in general having $\dot{x}$ as a constituent), then there is a type

$$[\dot{x} \mid s \models \sigma],$$

the *type* of all those objects x to which $\dot{x}$ may be anchored in the situation s, for which the conditions imposed by σ obtain in s.

This process of obtaining a type from a parameter $\dot{x}$, a situation s, and a compound infon σ, is known as *(object-) type abstraction*. The parameter $\dot{x}$ is known as the *abstraction parameter* used in this type abstraction.

The situation s is known as the *grounding* situation for the type. In many instances, the grounding situation s is 'the world' or 'the environment' we live in (generally denoted by w in my account). For example, the *type* of all people could be denoted by

$$[IND_1 \mid w \models \ll \text{person}, IND_1, l_w, i_{now}, 1 \gg]$$

Constraints, the facilitators and inhibitors of information flow, are abstract links between types of situation. They may be natural laws, conventions, logical (i.e., analytic) rules, linguistic rules, empirical, law-like correspondences, or whatever. Their role in the information chain is quite well conveyed by the use of the word *means*. For instance, consider the statement

smoke means fire.

This expresses a constraint (of the natural law variety). What it says is that there is a lawlike relation that links situations where there is smoke to situations where there is fire. If T_{smoke} is the type of situations where there is smoke present, and T_{fire} is the type of situations where there is a fire, then an agent (eg. a person) can pick up the information that there is a fire by observing that there is smoke (a type T_{smoke} situation) and being aware of, or *attuned to,* the constraint that links the two kinds of situation, denoted by

$$T_{smoke} \Rightarrow T_{fire}$$

(This is read as "T_{smoke} *involves* T_{fire}.")

As a constraint that holds 'because the world is that way', this constraint is one that holds regardless of the presence or absence of any cognitive agent, and is an instance of what is referred to as a *nomic constraint*. For an example of a non-nomic constraint, consider:

FIRE *means fire.*

This describes the linguistic constraint

$$T_{utterance} \Rightarrow T_{fire}$$

that links situations (of type $T_{utterance}$) where someone yells the word FIRE to situations (of type T_{fire}) where there is a fire. Awareness of this constraint involves knowing the meaning of the word FIRE and being familiar with the rules that govern the use of language.

The three types just introduced may be defined as follows:

$$T_{smoke} = [\dot{s} \mid \dot{s} \models \ll \text{smokey}, i, 1 \gg]$$
$$T_{fire} = [\dot{s} \mid \dot{s} \models \ll \text{firey}, i, 1 \gg]$$
$$T_{utterance} = [\dot{u} \mid \dot{u} \models \ll \text{speaking}, \dot{a}, i, 1 \gg \wedge \ll \text{utters}, \dot{a}, \text{FIRE}, i, 1 \gg]$$

The use of the same time parameter i in all three types indicates that there is no time-slippage in either case. The smoke and the fire are simultaneous, and so are the utterance and the fire. In general, parameters keep track of the various informational links that are instrumental in the way constraints operate.

Notice that constraints links types, not situations. On the other hand, any particular instance where a constraint is utilized to make an inference or modify behavior will involve specific situations (of the relevant types). Thus constraints function by relating various regularities or uniformities across actual situations.

4 The Sacks Analysis

I turn now to the examination Rosenberg and I made (Devlin and Rosenberg to appear) of Sacks' analysis of the two-sentence 'story opening':

The baby cried. The mommy picked it up.

The following object-types are particularly central to the analysis (i.e., to our situation-theoretic adaptation of Sacks' own analysis):

- $T_{baby} = [\dot{p} \mid w \models \ll \text{baby}, \dot{p}, t_{now}, 1 \gg]$, the type of all babies,

where $\dot{p}$ is a parameter for a person and w is 'the world', by which I mean any situation big enough to include everything under discussion;

- $T_{mother} = [\dot{p} \mid w \models \ll \text{mother}, \dot{p}, t_{now}, 1 \gg]$, the type of all mothers;
- T_{family}, a type of types, consisting of types such as 'baby', 'mommy', 'daddy', etc.;
- $T_{stage\text{-}of\text{-}life}$, a type of types, consisting of types such as 'baby', 'child', 'adult', etc.

The first two of the above types correspond to what Sacks refers to as *categories*. The correspondence is sufficiently close that I shall regard the terms 'category' and 'object-type (of individual)' as synonymous.

In addition to categories, Sacks also makes use of what he calls 'categorization devices'. He defines a *(membership) categorization device* (or simply a *device*) to be a non-empty collection of categories, together with rules of application. When endowed with the appropriate structure, the types T_{family} and $T_{stage\text{-}of\text{-}life}$ each correspond to a *categorization device*.

Categorization devices are structured objects, whose structure plays a role in the cognitive and social activities of agents that discriminate the devices and their component types.

For instance, the type $T_{stage\text{-}of\text{-}life}$ is linearly ordered:

$$T_{baby} < T_{child} < T_{adolescent} < T_{young\text{-}adult} < T_{adult} < T_{elderly}.$$

Of course, there are no clear-cut divisions between any adjacent pair of types in this chain. But when the vagueness of the types is 'factored out', the resulting order is indeed a linear ordering of the device.

People are aware of this ordering, and make use of it in their everyday lives. For instance, it can be used to praise or degrade others, using phrases such as 'acting like an adult' to praise a child, 'acting like a child' to degrade an adult, or 'acting like a baby' to degrade a child. In other words, people make systematic use of a relation

$$\text{less-than}(T, T')$$

between types T, T' in the device $T_{stage\text{-}of\text{-}life}$, a relation that, subject to vagueness of the types involved, is based on age.

What makes this kind of thing possible is that a society will often attach *values* to the nodes of the structure of a device. For instance, in contemporary American and European society, the optimal point in the stage-of-life device is usually T_{adult}, and progression along the device-ordering, starting from T_{baby}, and proceeding through T_{child}, $T_{adolescent}$, $T_{young\text{-}adult}$, and up to T_{adult}, is normally regarded as a steady increase in status in society, with the final step to $T_{elderly}$ regarded as a retrograde step down from the optimum.

The precise valuation may, however, vary from society to society. For example, in certain African tribes, status within the stage-of-life device increases throughout the entire linear ordering, with $T_{elderly}$ regarded as occupying the most-respected, optimal position in the device.

Academic hierarchies provide another, interesting example of a categorization device. For instance, if T_D is a State university-system, the device T_D has a precisely defined hierarchical structure (a tree) having a single top-node (the system president), directly beneath whom are a number of campus-presidents, together with the system-presidents' aids. Beneath each campus-president are the various deans along with the campus-presidents' aids, and so forth. Since the university system will be *defined* (as a university system) in this way, in this case the mathematical realization of the device-structure (a partial ordering) is an accurate model of the reality.

Such a hierarchy is constructed with the intention that the associated valuation is a monotonic one: the higher up the partial ordering one is, the greater is supposed to be the status, with the system-president occupying the most prestigious position of all.

Now, because of the highly visible and fundamental role this kind of structural hierarchy plays in the very establishment and operation of the system, those outside the system, such as journalists or government officials, will generally assume the intended valuation whenever they have dealings with the system. But other groups may attach other valuations to the structure, and indeed may add additional structure to the hierarchy. For instance, many research-oriented academics regard research professors or laboratory directors as occupying the optimal category, ranking well above administrative positions. In this case, the hierarchical structure is the same, what differs is the valuation associated with that structure.

Again, people familiar with the system often assume a linear or partial ranking of the individual institutions in the system, resulting in a corresponding ranking of the individual campus presidents, etc. This structure is additional to, and consistent with, the original system-hierarchy.

Of course, the provision of a mathematically precise structure does not at all amount to a claim that people do in fact act 'mathematically' (whatever that might mean). Rather the use of a mathematical structure simply provides us with a very precise descriptive device to handle one aspect of a highly complex phenomenon. Given such a formal conceptualization of a certain social structure as a mathematical structure with a social valuation, we may then go on to investigate the constraints that govern the way behavior is influenced by those features.

Other structural aspects of devices concern the constraints that link types within a device. I take up this issue next.

As originally developed by Barwise and Perry (1983), and used widely within situation theory and situation semantics (see (Devlin 1991)), constraints provide (are) links between situation-types, links that capture (are) systematic regularities connecting situations of one kind with situations of another. The idea is to provide a mechanism for studying context: the situations are for the most part contexts for the agent, contexts that influence the agent's activity.

Thus (Devlin 1991) provides the following 'standard' picture of constraints. A *constraint* is a particular kind of linkage $S \Rightarrow T$ that connects two situation-types S, T. Given a situation s of type S, the constraint $S \Rightarrow T$ provides the information that there is a situation t of type T. In typical cases, the situation t is, or may be regarded as, an extension of s, and thus the constraint provides information about the larger actuality of which s is part.

In order to address some of the questions Sacks investigates from the perspective of situation theory, it is necessary to extend the notion and theory of constraints quite a bit further. In particular, the notion of a constraint must be extended to cover not just situation-types but object-types as well.

In (Devlin and Rosenberg, to appear), we introduce constraints of the form $U \Rightarrow V$, where

$$U = [\dot{x} \mid u \models \sigma(\dot{x})] \quad , \quad V = [\dot{y} \mid v \models \tau(\dot{y})]$$

for some grounding situations u, v and compound (parametric) infons σ, τ. This constraint provides an 'associative link' between objects of type U and objects of type V, in that an agent A that is aware of, or attuned to, this constraint will, when it encounters an object X of type U *as an object of type U* and, under the same circumstances encounters an object Y of type V, will regard, or encounter, Y *as an object of type V*. (It is possible that X and Y are identical here, in which case we say that the constraint is functioning *reflexively*.)

For example, there is a constraint

$$T_{baby} \Rightarrow T_{mother}$$

such that, if $\mathcal{A}$ is attuned to this constraint and encounters an object B of type T_{baby} and an object M of type T_{mother} , and furthermore, $\mathcal{A}$ encounters B *as an object of type* T_{baby}, then $\mathcal{A}$ will in fact encounter M as an object of type T_{mother}.

In (Devlin and Rosenberg, to appear), we call such a constraint an *associative link*. Associative links are an important special case of what we call *normative constraints*; these are informational constraints that guide the normal behavior of agents.

Of course, though an associative link might provide a situation-theoretic mechanism corresponding to normal behavior, it does not on its own say anything about what evokes a particular link on a given occasion. Indeed, the same may also be said of the more familiar situation-type constraints described in (Devlin 1991): what makes a particular constraint *salient* on a given occasion, so that an agent does in fact modify its behavior in response to that constraint? The answer is, in large part, beyond the scope of situation theory, which sets out to provide a framework for studying cognitive behavior in terms of information flow 'in the world' (i.e., between agents), but does not attempt to provide a theory of mind. However, part of the answer is at the very least closely connected to the issue of just what it is in the world that *supports* a given constraint, the issue I take up next.

Let me start with the more familiar kinds of constraints that link situation-types.[7] To be even more definite, let me take the following oft-used examples of constraints discussed in (Devlin 1991: 91–94):

(C_1)   *smoke means fire*

(C_2)   *the ringing bell means class is over*

(C_3)   COOKIE *means cookie*

C_1 is what is normally referred to as a *nomic* constraint, one that obtains because the world is the way it is. Now, C_1 is a fact about the world, and as such may be represented as an infon, namely

$$\sigma_1 = \ll \text{involves}, T_{smoke}, T_{fire}, l_C, t_C, 1 \gg,$$

where l_C and t_C are the location and time periods in which this constraint obtains. (In particular, l_C includes the surface of the Earth and t_C includes the present epoch.) In fact, according to the situation-theoretic view of the world, when considered as a fact about the world, the constraint C_1 *is* this infon, and

---

[7] The following discussion is taken almost verbatim from (Devlin and Rosenberg to appear). I include it here for the convenience of readers who do not have a copy of that paper to hand.

the factuality of C_1 is expressed by the proposition that σ_1 is supported by the appropriate situation.

The question now is, what is the appropriate situation? It is consistent with the level to which the theory was developed there to take the supporting situation to be the whole world, w, and express the factuality of the constraint C_1 by means of the proposition

$$w \models \sigma_1.$$

For the present analysis, however, such an answer is far too crude. We need to ask ourselves just what it means to say a certain constraint obtains.

Well, the constraint C_1 is a certain regularity in the world, an abstract feature of the world comprising a systematic linkage between situations of the two types T_{smoke} and T_{fire}. But what is the ontological status of this systematic linkage other than a situation? (See (Devlin 1991: 69-85) for a discussion of the nature of situations.) And it is precisely this situation that supports C_1, of course. I shall denote this situation by $Support(C_1)$.

In general, any nomic constraint C will comprise a systematic, informational link between pairs of situations, situations of types T and T', respectively, and as a systematic regularity in the world, this linkage will constitute a situation, which I shall denote by $Support(C)$. This situation will include, in particular, the relevant causality between situations of type T and those situations of type T' to which they are linked. Moreover,

$$Support(C) \models C$$

and $Support(C)$ is in a sense the 'minimal' situation that supports C.

Of course, there is a certain apparent triviality to the above, in that I seem to come close to claiming that it is the constraint that supports the constraint. Well, in a way I am, but the matter is not quite as simple as that, and besides even this *apparent* triviality is not the case for all constraints, and indeed is, in general, characteristic of the nomic constraints. A nomic constraint holds in the world because that is the way the world is; what makes a particular regularity a (nomic) *constraint* is the role it plays in guiding the flow of information. That is to say, in the case of a nomic constraint, the distinction between the regularity (i.e., situation) and the constraint is essentially one of abstraction and functionality.

Thus, an agent that observes instances of smoke being related to fire observes parts of a certain situation, $Support(C_1)$. As a result of a number of such observations, perhaps backed up by other means of acquiring information about smoke and fire, the agent might then come to be aware of, or attuned to, the constraint C_1 (abstraction), and to make use of that constraint (functionality). But the agent is not going to become aware of the *situation* $Support(C_1)$; such highly abstract situations are not the kind of object that agents normally become aware of (as entities in the world).

For a nomic constraint such as C_1, then, it is the very uniformity linking smokey situations to firey situations that constitutes the situation that supports the constraint, and the constraint may be regarded as an *abstraction* from that situation. Turning to the second of the three examples, C_2, things are quite different. Here it is not at all the systematic linkage between situations in which a bell rings and situations in which the class is over that constitutes the supporting situation. The connection between two such situations is not, after all, an aspect of the way the world is, even if we restrict things to the particular educational establishment concerned. Rather the connection exists purely by virtue of some previously established convention. That is to say, the 'minimal' situation that supports C_2, $Support(C_2)$, the situation *by virtue of which* C_2 obtains, is the situation comprising the convention. This is, of course, a highly abstract situation that exists purely as an aspect of the behavior of a certain community. As far as I know, prior to (Devlin and Rosenberg, to appear), situations of this kind had not previously been considered in situation theory. Though they do not readily fall into the category of 'parts of the world' that is often used to explain the concept of a situation, they do arise in a quite natural way when we ask the question "What is it in the world that supports a particular constraint (i.e., infon)?" I shall have more to say on the nature of situations in just a moment.

Of course, a particular student may come to be aware of the constraint C_2 by observing the systematic way ringing bells are followed by the ending of class, without ever being informed that there is a previously established convention to this effect, but this says nothing about what it is in the world that actually *supports* this constraint.

My third example, C_3, is a linguistic constraint connected with the (American) English language. Though one could take the supporting situation to be the regularity that connects uses of the word COOKIE to instances of cookie (i.e., to actual cookies), this would amount to reducing language to a whole bag of isolated words and connections. More reasonable is to take the supporting situation in the case of a linguistic constraint to be the entire language situation; that is to say, the complete collection of rules, norms, and practices that constitutes the English language. Again this is a highly abstract kind of situation, not previously encountered in situation theory (at least not in the form of a situation in the ontology). It is similar in nature to the 'ringing bell means class is over' constraint just discussed, except that the ringing bell support-situation is restricted to one particular constraint whereas the English language situation supports all English meaning relations.

Notice that we have now arrived at a much more general notion of 'situation' than the one explicitly described in (Devlin 1991) (though that treatment simply left open the possibility of extending the notion along the present lines). We may think of situations as aggregates of features of the world that constitute an *aspect* of the world. That aspect may be physical, possibly located in time and/or space, or abstract, perhaps consisting of systematic regularities in the world, as in the case of nomic constraints, or maybe arising from the intentional activity of cognitive agents, as in the case of language or social conventions.

With the discussion of the more familiar kind of constraint behind us now, I turn to the associative links. Consider first the constraint

$$T_{baby} \Rightarrow T_{mother}.$$

This links the two concepts of babies and mothers. What supports it? Well, it is one of the constraints that could be said to be part of the 'world of families'. But in situation-theoretic terms, the 'world of families' constitutes a situation. (More precisely, it constitutes a number of situations, the 'close family' situation and various 'extended family' situations. I shall restrict my attention to the former for the time being.) This situation comprises everything that is part and parcel of being a family, and thus is captured within the existing situation-theoretic framework as $Oracle\,(T_{family})$. (More precisely, as $Oracle_\Gamma(T_{family})$, for an appropriate set of issues Γ. See (Devlin 1991: Section 3.5) for details.) This is the situation that supports the constraint:

$$Oracle\,(T_{family}) \models T_{baby} \Rightarrow T_{mother}.$$

Notice that for an associative link such as this, it is not so much the validity of this proposition that is relevant, as is the issue of whether or not the link is *salient* to the agent at a particular juncture. That is to say, what counts is whether the agent's behavior is governed by, or guided by, the link. And the agent's behavior will be so influenced if it is currently operating in the context $Oracle\,(T_{family})$. I shall write

$$Oracle\,(T_{family}) \,\|\!-\, T_{baby} \Rightarrow T_{mother}$$

to indicate that the constraint $T_{baby} \Rightarrow T_{mother}$ is *salient* in $Oracle\,(T_{family})$.

The notation $s \,\|\!-\, C$ is borrowed from set theory, where it is articulated "*s forces C*".

In general, a given situation will support a great many constraints, many of them irrelevant as far as the agent's current activity is concerned. But for many situations, in particular the ones that arise as contexts for cognitive activity, there is often a 'hierarchy' of constraints, with some constraints being more prevalent (i.e., more likely to influence the agent's behavior) than others. I refer to the prevalence of a particular constraint or constraints in a situation as *salience*, and write

$$s \,\|\!-\, C$$

to indicate that the constraint C is *salient* in the situation s. Clearly,

$$s \,\|\!-\, C \quad \text{implies} \quad s \models C$$

but not conversely.

When an agent acts within a particular context s, its behavior will be strongly influenced by the constraints that are salient to s (among those supported by s). The greater the salience, the more will be the influence, though at this level of analysis I shall not introduce any non-trivial spectrum of salience.

The major complicating factor here is that, in general, an agent will be simultaneously operating in a number of contextual situations, each bringing with it its own salience hierarchy, and the different constraints that are salient in the respective contexts may well 'compete' for influence. Sometimes two constraints will be mutually supportive, but on other occasions salient constraints in two contexts may be mutually contradictory, and the agent must 'choose' between them. How all of this works, or rather what mechanisms may be developed to provide a useful description of this phenomenon, is an issue that needs to be taken up by situation-theorists in due course.

The salience of a particular constraint C in a situation s is often a reflection of the fact that C is closely bound up with the structure of s.

For example, the situation $Oracle\,(T_{family})$ supports many constraints concerning family relationships and the way members of a family behave to each other, including constraints that, while they undoubtedly guide and influence an agent's behavior, do so in a highly peripheral way as far as purposeful activity is concerned, such as the constraint that a family member is human, that husbands are male, etc. On the other hand, some constraints are particularly salient to this situation, in particular those that concern the family relationships that bind the family together as a family, such as the associative link that says a mother is the mother of a child, a husband is the husband of a wife, etc.

In particular, the type $T_{mother\text{-}of}$ acts as a 'fundamental' one within T_{family}, with the types T_{mother} and T_{baby} being linked to, and potentially derivative on, that type. More precisely, the following structural constraints are salient in the device T_{family}:

$$T_{mother} \Rightarrow \exists \dot{y} T_{mother\text{-}of}$$

$$T_{baby} \Rightarrow \exists \dot{x} T_{mother\text{-}of}$$

where T_{mother} is the 2-type

$$[\dot{x}, \dot{y} \mid w \models \ll \text{mother-of}, \dot{x}, \dot{y}, t_{now}, 1 \gg].$$

This has the following consequence: in the case where $T_{mother} : T_{family}$ (i.e., T_{mother} is of type T_{family}) and $T_{baby} : T_{family}$, the following implications are salient:

$$p : T_{mother} \ \longrightarrow \ \exists q\,(p, q : T_{mother\text{-}of}) \tag{1}$$

$$q : T_{baby} \ \longrightarrow \ \exists p\,(p, q : T_{mother\text{-}of}). \tag{2}$$

These two implications are not constraints. In fact they do not have any formal significance in situation theory. They are purely guides to the reader as to where this is all leading.

It should be noted that normative constraints, such as the associative links, are a product of evolution: they evolve along with the society they pertain to. The result of such an evolutionary process is a sort of 'chicken-and-egg' state of affairs, whereby members[8] can be said to behave *according to certain normative constraints*, and indeed on occasions may even explicitly *follow* some of those

---

[8] I generally use the word 'member' in the sociologist's sense, to mean 'member of a society', or, depending on context, 'member of the society under consideration', and use the word 'agent' for the more general notion of any cognitive agent, not necessarily human.

constraints, and yet those constraints obtain *by virtue of the fact that members normally behave that way.* A similar remark can be made about linguistic constraints, of course.

Category-binding constraints constitute another class of constraints that arise from a situation-theoretic analysis of the Sacks paper.[9] A *category-binding constraint* is a link between an activity (i.e., a relation in the ontology) A and a type T. The salience of such a constraint means that a person would normally associate the activity A with objects of type T.

As in the case of associative links, the issue of relevance for a category-binding constraint is salience in a context rather than factuality (i.e., being supported by the context). For instance, in the device $T_{stage\text{-}of\text{-}life}$, the activity of crying is *category bound* to the type T_{baby}; thus

$$Oracle\,(T_{stage\text{-}of\text{-}life})\ \|\!\!-\ crying \Rightarrow T_{baby}.$$

According to Sacks, evidence that the activity of 'crying' is category-bound to the type T_{baby} in the device $T_{stage\text{-}of\text{-}life}$ is provided by the fact (mentioned earlier) that members recognize a link between the activity of 'crying' and the position of the category T_{baby} in the linear-ordering structure of the device. For instance, a member of T_{child} who cries might be said to be 'acting like a baby' and a member of T_{baby} who does not cry when the circumstances might warrant it, such as following an injury, might be praised as 'acting like a big boy/girl'.

This seems to be an instance of a more general phenomenon, whereby one feature of an activity A being category-bound to a type T_C in a device T_D is that A is closely related to the position of T_C within the structure on T_D.

When category-binding arises, as it generally does, by virtue of the very structure of a particular device, listeners can make use of this tight relationship to obtain what, in a slightly different context, Barwise and Perry (1983: 167–168) have called 'inverse information'.

For example, utterance of the word 'baby' can trigger in the listener a number of different information structures, among them the device $T_{stage\text{-}of\text{-}life}$ and the device T_{family}. Each of these devices will bring to bear different sets of constraints, which will influence subsequent processing in different ways. Hearing the sentence "The baby cried", however, the combination of 'baby' with 'crying' can suggest to the listener that the appropriate device here is $T_{stage\text{-}of\text{-}life}$. Thus, the listener will normally hear the utterance of this sentence in such a way that

$$T_{baby} : T_{stage\text{-}of\text{-}life}.$$

That is to say, this item of information will be available to the listener as he processes the incoming utterance, and will influence the way the input is interpreted.

Since the normal situation-theoretic scenario is for an available context to combine with an utterance to determine information content, the inference of

---

[9] Though Rosenberg and I did not single out these particular constraints in our paper (Devlin and Rosenberg, to appear).

an appropriate context from an utterance (together with other, background information, such as social knowledge) in the way just considered is an instance of 'inverse information'. In his paper, Sacks presents other examples of this phenomenon that arise in everyday discourse. Indeed, as Sacks observes,[10] people often make use of the device of category-binding and inverse information to avoid direct mention of delicate or unpleasant contextual features.

Pulling the various observations together now, here is the situation-theoretic analogue of Sacks' analysis of the two-sentence data.

Let s denote the described situation for the discourse. The situation s will be such that it involves one and only one baby, otherwise the use of the phrase 'the baby' would not be appropriate. In starting a communicative act with the sentence "The baby cried", the speaker is informing the listener that she is commencing a description of a situation, s, in which there is exactly one baby, call it b.

The *propositional content* of the utterance of the first sentence "The baby cried", which in the case of a simple description like the present example is, for speaker and listener, the principal item of information about the described situation that is conveyed, is

$$s \models \ll \text{cries}, b, t_0, 1 \gg$$

where t_0 is the time, prior to the time of utterance, at which the crying took place. In words, in the situation s, the baby b was crying at the time t_0.

Notice that, in the absence of any additional information, the only means available to the listener to identify b is as the referent for the utterance of the phrase 'the baby'. The utterance of this phrase tells the listener two pertinent things about s and b:

$$b : T_{baby} \quad \text{(i.e., } b \text{ is of type } T_{baby}) \tag{3}$$

where T_{baby} is the type of all babies, and

$$b \text{ is the unique individual of this type in } s. \tag{4}$$

As noted earlier in the discussion of category binding, in addition to the propositional content, the utterance of the entire first sentence "The baby cried" also provides the listener with the information

$$T_{baby} : T_{stage\text{-}of\text{-}life}. \tag{5}$$

The speaker then goes on to utter the sentence "The mommy picked it up." As in the case of 'the baby', in order for the speaker to make appropriate and informative use of the phrase 'the mommy', the described situation s must contain exactly one individual m who is a mother. In fact we can make a stronger claim: the individual m is the mother of the baby b referred to in the first sentence. For if m were the mother not of b but of some other baby, then the appropriate

---

[10] Though using a different theoretical framework.

form of reference would be 'a mother', even in the case were m was the unique mother in s. The mechanism that produces this interpretation can be described as follows.

Having heard the phrase 'the baby' in the first sentence and 'the mommy' in the second, the following two items of information are salient to the listener:

$$m : T_{mother} \qquad (6)$$

$$m \text{ is the unique individual of this type in } s. \qquad (7)$$

In addition, I shall show that the following, third item of information is also salient:

$$m \text{ is the mother of } b. \qquad (8)$$

Now, as I have already observed, because the activity of *crying* is category-bound to the type T_{baby} in the device $T_{stage\text{-}of\text{-}life}$, following the utterance of the first sentence, the listener's cognitive state is such that the type T_{baby} is of type $T_{stage\text{-}of\text{-}life}$. This type has categories that include T_{baby}, T_{child}, $T_{adolescent}$, T_{adult}, all of which have equal ontological status within this device, with none being derivative on any other. But as soon as the phrase 'the mommy' is heard, the combination of 'baby' and 'mommy' switches the emphasis from the type $T_{stage\text{-}of\text{-}life}$ to the type T_{family}, making salient the following propositions:

$$T_{baby} : T_{family}. \qquad (9)$$

$$T_{mommy} : T_{family}. \qquad (10)$$

In the T_{family} device, the various family relationships that bind a family together (and which therefore serve to give this device its status as a device) are more fundamental than the categories they give rise to. In particular, the types T_{baby} and T_{mother} are derivative on the type $T_{mother\text{-}of}$ that relates mothers to their babies.

Now, proposition (9) is the precondition for the salience of implication (2), namely

$$q : T_{baby} \ \rightarrow\ \exists p \, (p, q : T_{mother\text{-}of}).$$

Substituting the particular individual b for the variable q, we get

$$b : T_{baby} \ \rightarrow\ \exists p \, (p, b : T_{mother\text{-}of}).$$

But by (3), we know that

$$b : T_{baby}.$$

Thus we have the salient information

$$\text{there is an } m \text{ such that } m, b : T_{mother\text{-}of}. \qquad (11)$$

The use of the definite article in the phrase 'the mommy' then makes it natural to take this phrase to refer to the unique m that satisfies (11). Thus the listener naturally takes the phrase 'the mommy' to refer to the baby's mother. This interpretation is reinforced by the completion of the second sentence "...picked it up", since there is a social norm to the effect that a mother picks up and comforts her crying baby. This explains how the fact (8) becomes salient to the listener.

It should be noticed that the switch from the salience of one set of constraints to another was caused by the second level of types in the type-hierarchy. The constraints we were primarily interested in concerned the types T_{mother} and T_{baby}. These types are part of a complex network of inter-relationships (constraints). Just which constraints in this network are salient to the agent is governed by the way the agent encounters the types, that is to say, by *the type(s) of those types*— for instance, whether T_{baby} is regarded (or encountered) as of type $T_{stage-of-life}$ or of type T_{family}. This consideration of the type of the types linked by a constraint represents a much finer treatment of constraints and inference than has hitherto arisen in situation theory. By moving to a second level of typing (i.e., to types of types), we are able to track the way agents use one set of constraints rather than another, and switch from one set to another. The first level of types allows us to capture the informational connections between two objects; the second level allows us to capture the agent's preference of a particular informational connection, and thereby provides a formal mechanism for describing normality.

5   Normality

The informational structure that the situation-theoretic framework, and in particular the type structure, brings to the analysis of the Sacks data results in our avoiding a number of problems that, as we point out in (Devlin and Rosenberg, to appear), Sacks' own argument runs into.

Of the several points in Sacks' analysis that our reformulation in situation-theoretic terms brings to light as in need of clarification, by far the biggest problem we encountered concerns Sacks' *Hearer's Maxim 2*:[11]

> (HM2) If some population has been categorized by means of a duplica-
> tively organized device, and a member is presented with a categorized
> population which *can* be heard as 'coincumbents' of a case of that de-
> vice's unit, then hear it that way.

As we point out in (Devlin and Rosenberg, to appear), this maxim is logically inconsistent. (The inconsistency shows up dramatically as soon as one tries to recast Sacks' argument in a more mathematical fashion.) The problem lies in the wide scope that results from the use of the word 'can'. The *possibility* of hearing something a certain way does not mean that one 'should' hear it that way: possibility does not imply normality.[12]

---

[11] Sacks, notion of a 'duplicatively organized device', of which the family device is an instance, is somewhat complicated to explain. Since it plays no role in the argument of the present paper, I shall refer readers to (Sacks 1972) or (Devlin and Rosenberg, to appear) for details.

[12] It should be pointed out that this has nothing to do with giving 'possibility' *too* wide a scope; the inconsistency is present even if one adopts the reading, as Sacks surely intended, of 'possibility within the given context.'

The 'fix' we suggest in (Devlin and Rosenberg, to appear) is to replace Hearer's Maxim 2 by the alternative:

(HM2') If some population has been categorized by means of a duplicatively organized device, and a member is presented with a categorized population which would *normally* be heard as coincumbents of a case of that device's unit, then hear it that way. [Emphasis added.]

As we remark in (Devlin and Rosenberg, to appear), from a logical point of view, this appears to verge on the tautologically trivial. But, as we also remarked, it does not appear to be vacuous. Indeed, from the point of view of coming to grips with the issue of normality, we believe it has significant content. Not only does the act of *stating* (HM2') bring the concept of normality into the analysis, indeed as a crucial feature of that analysis, the maxim itself captures the essence of normality with regards to the way people understand certain utterances. (Notice that if you replace 'normally' with some other qualifier in the statement of (HM2'), the resulting maxim is unlikely to be a reliable observation of, or norm for, everyday human behavior.)

The same problem arises for Sacks' *Hearer's Maxim 1*:

(HM1) If two or more categories are used to categorize two or more members of some population, and those categories *can* be heard as categories from the same device, then hear them that way. [Emphasis added.]

Again, the wide scope given by the word 'can' makes this maxim is inconsistent. And once again, my suggestion would be to modify this maxim as follows:

(HM1') If two or more categories are used to categorize two or more members of some population, and those categories would *normally* be heard as categories from the same device, then hear them that way.

We did not dwell on the issue of the inconsistency of (HM1) in (Devlin and Rosenberg, to appear), since this particular maxim was not required for Sacks' final analysis of the two-sentence data, which was our focus of attention at the time.

As mentioned above, the problem with (HM2) only showed up when we were trying to reformulate Sack's argument in situation-theoretic terms. In the situation-theoretic version of the argument, the problem does not arise. Tracking the information flow involved in a 'normal' understanding of Sacks' two sentences, the understanding that (HM2) is intended to lead to (or reflect) falls straight out of the analysis. Our reason for formulating (HM2') was not that it was needed in the situation-theoretic analysis. Rather, (HM2') was our suggestion, *based upon* our situation-theoretic analysis, of the change that needed to be made to (HM2) *in order for Sacks' own argument to work*.

Of course, if Sacks' analysis depends upon normality in a critical fashion, as we suggest, then normality must surely play a role in our own analysis. This is indeed the case, but in a somewhat different guise. In the situation-theoretic framework, the procedural or behavioral notion of *normality* is replaced by the structural notion of *salience*.

In situation-theoretic terms, *normal behavior* in a particular context amounts to acting in accordance with the constraints that are most salient in that context. Thus an investigation of the concept of normality is reduced (or transferred, to use a less value-laden term) to a study of salience, a topic whose importance we stressed in (Devlin and Rosenberg, to appear), but a detailed examination of which was deferred until later.

The apparent triviality of (HM2′) as a maxim is a consequence of (HM2′) being, in effect, a *definition* (or, if you prefer, a *classification*) of normality: "normally do what you would normally do". By moving from a procedural notion to a structural one (salience), this apparent circularity is avoided. More precisely, the vicious circularity of the notion of normality is replaced by the recursive procedure whereby agents use constraints to which they are attuned, or of which they are aware, in order to develop an attunement to, or an awareness of, further constraints. This is the phenomenon I address in the next section.

6 Constraint Formation and Salience

The main function of the situation-theoretic notion of *type-abstraction* is to capture, within the theory, the ability, characteristic of cognitive agents (see (Dretske 1981: 142)) and a propensity in the case of human agents, to recognize, and on occasion to create, similarities between various things in the world and to construct categories (or 'types') of things according to those similarities. It is indeed by virtue of the systematization of an agent's environment provided by means of such typing that the agent is able to behave in what we generally refer to as a 'rational' (as opposed to a random or haphazard) manner.

Thus, for example, people recognize the categories of *baby, mother, family,* and *stage-of-life*. In situation-theoretic notation, these types may be obtained by means of the following type-abstractions:

- $T_{baby} = [\dot{p} \mid w \models \ll \text{baby}, \dot{p}, t_{now}, 1 \gg]$;
- $T_{mother} = [\dot{p} \mid w \models \ll \text{mother}, \dot{p}, t_{now}, 1 \gg]$;
- $T_{family} = [\dot{e} \mid w \models \ll \text{family-type}, \dot{e}, 1 \gg]$;
- $T_{stage-of-life} = [\dot{e} \mid w \models \ll \text{stage-of-life-type}, \dot{e}, 1 \gg]$;

where $\dot{p}$ is a parameter for a person, w is 'the world', i.e., any situation big enough to include everything under discussion, and $\dot{e}$ is a parameter for a type.

Now, each of these identities provides a *structural* dependency of the type on the corresponding property, with the *type* T_{baby} the result of *abstracting* across the *property* 'baby', etc. But this is a theoretical dependency. We could equally well provide a mechanism for recovering the property from the type, if we needed such a device. Maintaining a distinction between properties and types is convenient for the purposes of developing a theory of information, but in terms of the target domain of cognitive agents, properties and types are very much two sides of the same coin, and moreover two sides that look very much alike. (The principal technical difference is that types incorporate the notion of 'truth in a situation' whereas properties are just that, properties, requiring a context in order to encompass truth.)[13]

Nevertheless, despite the somewhat artificial nature of the technical distinction between properties and types, types do help us capture, in a natural way, the notions of *category* and *categorization device* that Sacks uses in his analysis.

The question I want to investigate now is, what does *recognition* (by a person) of a category (i.e., type) amount to, and how does a person *acquire* these categories and the ability to use them (i.e., the ability to place various entities in the appropriate category)?

In the case of *categories* (types of individuals, in situation-theoretic terms), things seem fairly straightforward. Recognition of, or familiarity with, a category (or type) is simply a matter of recognizing those individuals that are in the category (are of the type) and those that are not.[14] Thus, familiarity with the 'baby' category (type T_{baby}) is a matter of recognizing babies (that is to say, recognizing certain individuals *as babies*), and likewise familiarity with the 'mother' category (type T_{mother}) is a matter of knowing what a mother is.

Moreover, *use* of a category in everyday life, including cognition and communication, requires, for the most part, nothing more than the ability to recognize individuals as being of that category. Once I know what a 'baby' is, I can understand references to babies and can myself make reference to babies.

It is at the level of *categorization devices* (types of types) that things get a bit more interesting.

Knowing what a 'family' or a 'stage-of-life' is, and being able to recognize that a certain category is a family category or a stage-of-life category, involves quite a bit more than the ability to recognize an individual as being, or not being, in a certain category (of a certain type). This is in part because of the increased level of abstraction: categorization devices (types of types of individuals) are at a more abstract level than categories (types of individuals). But the most significant difference is that categorization devices have a *structure* that is important *to their being categorization devices*, whereas categories do not.

For instance, familiarity with the family device involves an awareness that a family is not just a collection of two or more persons (or animals of some species), but a collection bound together by certain kinds of relationships, relationships involving geneology and/or certain kinds of legal commitments, and possibly less institutionalized social commitments. That is to say, one cannot know what

---

[13] For an example where the distinction is fairly dramatic, consider the property of 'unicorn-ness', i.e. the property of being a unicorn. This is a meaningful, *bona fide* property, closely related to the properties of being a horse and having a horn. It is a property that figures in many stories, and is part of human culture. However, when abstracted over w, the world, this property leads to the vacuous type, which is only a type in a trivial, degenerate way.

[14] I am speaking here purely of recognition capacity. The issue as to *how*, on a given occasion, a member 'chooses' a particular way of regarding an individual in preference to other, equally valid type-ascriptions, is one I shall take up presently.

a 'family' is (in the sense of having the 'family' categorization device at one's cognitive disposal) without knowing (at least some of) the kinds of relationships that bind families together. (Note that, in (Sacks 1972), Sacks' explicitly includes 'rules of application' in his definition of a categorization device.)

Likewise, familiarity with the stage-of-life device entails knowing at least some of the categories in this device, together with the ordering relation between them, since it is the ordering relation between the various categories that makes this particular device the *stage-of-life* device.

The structure that is an essential part of a categorization device played a significant role in Sacks' analysis of the two-sentence data, and an even greater role in the alternative analysis given by Rosenberg and myself, where the structure associated with (indeed, constitutive of) the family device enabled us to circumvent Sacks' problematical (HM2) maxim.

Now, in (Devlin and Rosenberg, to appear), we did not explicitly introduce a situation-theoretic analogue of a categorization device. Instead, we carried out our argument using the types T_{family} and $T_{stage-of-life}$, plus the oracles of these types, $Oracle(T_{family})$ and $Oracle(T_{stage-of-life})$. It is these latter two, oracle situations that most closely correspond to Sacks' 'family' and 'stage-of-life' categorization devices. Indeed, this is how we implicitly regarded them when we were working on (Devlin and Rosenberg, to appear). We chose to make use of the oracle operator of situation theory, rather than introduce categorization devices as a further entity in the situation-theoretic ontology, for reasons of ontological parsimony. For the level of treatment we are currently able to give for data such as that of Sacks, the oracle situations corresponding to types seem to fulfill our aims. However, we do not rule out the necessity of having to abandon this technical convenience at some future date, and develop a separate situation-theoretic notion of 'categorization device'.

As I mentioned earlier, and as Rosenberg and I argued in (Devlin and Rosenberg, to appear), the pivotal role played by the family-binding relationships in the family device (i.e., in that device *being* a device), is reflected in the fact that the corresponding constraints are *salient* in $Oracle(T_{family})$. Similarly, the various constraints concerning the ordering relation between the various categories in the stage-of-life device are all salient in $Oracle(T_{stage-of-life})$.

Since the constraints that are salient in a particular device constitute an integral part of that device, a member's ability to utilize a categorization device entails an awareness of, or attunement to, the constraints that are salient in that device, and the process of achieving that ability involves acquiring that awareness or attunement. The process whereby a member acquires the ability to utilize a device can be—and, I believe, usually is—a mixture of being told, or otherwise informed, about the device, repeated experience in encountering the device, and perhaps conscious inner reflection on those experiences.

According to the account Rosenberg and I are currently developing, *normal behavior* in a given context amounts to behavior in accordance with the constraints salient (or most salient) in that context. In order to fully appreciate what this means, I need to say something about salience. What I say will, of necessity, be somewhat tentative, since Rosenberg and I have not yet fully worked out what we take to be a satisfactory account of this notion.

'Rational' behavior by an agent amounts to that agent acting and reasoning in accordance with various constraints, constraints that obtain in the context in which that agent is situated. But a given situation will, in general, support a great many constraints, many of them quite possibly of a minor and, to all intents and purposes, irrelevant nature as far as the agent's activity is concerned. This is where the salience 'hierarchy' comes in. The more salient is a constraint C within a given supporting situation s, the more likely is an agent for whom s is a context to be guided by C.

For example, in real-world, environmental situations, the constraint that links smoke to fire is a highly important one, crucial to the agent's survival, and as such is, under normal circumstances, likely to be the most salient in any situation in which there is smoke. On the other hand, one can imagine that a group of people has established a constraint whereby a smoke-signal indicates that the enemy is approaching, and for the members of that group, their context is such that the most salient constraint is the one that links smoke to an approaching enemy. Since the behavior of an agent is most likely to be influenced by the most salient constraints in its current context, different contexts can bring different constraints into play.

A metaphor I find particularly helpful in connection with salience is that of 'living in a situation'. Situations are an attempt to capture, within a rigorously defined ontology, the role played by context in reasoning and communication. Now the key feature of context—what makes it a *context*—is that it guides the flow of information *from without*, it does not play an explicit role. For example, in the case of everyday activity, the way you behave in the home situation differs significantly from your behavior at work, and is different again from the way you act when in a concert hall. Each situation carries with it a distinctive set of rules and norms of behavior, which situation theory treats as constraints. Again, in the case of communication, you make use of different linguistic constraints, depending on whether you are speaking to a small child, to your family, or at work, and different constraints again when speaking a foreign language. Now in each of these cases, you do not make explicit use of any *rules* (except perhaps on very special occasions which need not concern us here); the context does not play an active role in your activity, it simply *establishes the appropriate constraints*. During the course of any particular situated activity, you may be thought of as 'living in the world' that constitutes the context for that activity.

Notice that there is no reason to suppose you are restricted to 'living in' just one situation at a time in this manner. The high degree of abstraction allowed for situations enables us to use the theory to capture many different aspects of behavior at any one time. When playing a game of cards with some friends, for example, your behavior will be guided by the formally specified rules of the game, the linguistic rules governing English language, and the social rules and norms that guide group behavior when among friends. According to the 'living in a situation' metaphor, you may thus be described as simultaneously living in three different situations.

Similarly, to be competent in a particular language amounts to having a familiarity with a certain abstract situation, to 'know one's way around that situation'. Speaking or understanding a particular language can be regarded as 'cognitively living in' the appropriate linguistic situation, in the same way that finding one's way around one's home can be regarded as physically living in that home situation. Learning a new language involves becoming familiar with another linguistic situation, learning to find one's way around that situation.

Of course, salience is not an all-or-nothing state of affairs. Preferential readings, interpretations, or actions can be over-ruled or ignored. Another analogy that I find helpful is to think of salience as analogous to a gravitational field. At any point, the field pulls you in one particular direction, and depending on your circumstances you will follow that pull to a greater or lesser degree. If there are no other forces (constraints, influences) acting on you, you will follow the field in the direction of strongest pull—which itself may well be the resultant of component attractions in a number of different directions. When there are other forces acting on you, you will still be under the influence of the field, but may move in a different direction, a direction that results from the combined effect of the gravitational field and those other forces. *Normality* can be thought of as following the attraction of a 'gravitational field'.

Associative links, which Rosenberg and I introduced in (Devlin and Rosenberg, to appear), are one instance of a class of constraints that reflect/guide normal behavior. The associative link that connects the baby category to the mother category is salient in the family device, i.e.,:

$$Oracle\,(T_{family})\ ||\!\!-\ T_{baby} \Rightarrow T_{mother}$$

Now, as we claimed in (Devlin and Rosenberg, to appear), associative links are a quite new kind of constraint in situation theory. However, when we ask ourselves how a member would normally become attuned to, or aware of, a constraint such as the one above, we are at once thrown back to the more familiar notion of a situation-theoretic constraint between (environmental) situation-types.

Attunement to the constraint

$$T_{baby} \Rightarrow T_{mother}$$

would normally develop as part of the process of becoming familiar with the 'family' device, $Oracle\,(T_{family})$. The member (presumably a small child) repeatedly encounters situations s in which there is close interaction between a mother and her baby. As a result of these encounters, the member comes to recognize a certain *type* of situation, the type of situation in which there is such interaction. Let T_{bm} denote this situation-type.

Types such as T_{bm} are what I propose to call *evolved* types. Such types cannot normally be defined in an extensional fashion, and thus may not be effectively specified by means of the normal mechanisms for type-abstraction within situation theory. They arise by a process of repeated interaction between certain kinds of agent and the environment. Such agents include people, various species of animal, and neural networks.

In the case of the evolved type T_{bm}, one aspect of particular interest to us at the moment is that it involves two individuals, a baby and a mother. Let $\dot{b}$ be a parameter for a baby, $\dot{m}$ a parameter for a mother. I shall write $T_{bm}(\dot{b}, \dot{m})$ to indicate that the type T_{bm} has $\dot{b}$ and $\dot{m}$ as constituents.

As part of the process of *evolving* the type T_{bm} (i.e., acquiring the ability to discriminate this type), a member will learn that, under such circumstances, the two major participants normally consist of a mother and her baby. This amounts to developing an attunement to the constraint

$$T_{bm}(\dot{b}, \dot{m}) \Rightarrow T_{is}$$

where

$$T_{is} = [\dot{s} \mid \dot{s} \models \ll \text{baby}, \dot{b}, t_{now}, 1 \gg \\ \wedge \ll \text{mother}, \dot{m}, t_{now}, 1 \gg \\ \wedge \ll \text{mother-of}, \dot{m}, \dot{b}, t_{now}, 1 \gg]$$

A member who is attuned to this constraint, and who encounters a situation that is recognizably of type T_{bm}, will be aware that the individuals to which the parameters $\dot{m}$ and $\dot{b}$ are anchored will normally consist of a mother and her baby.

It is natural to take the support of the above constraint to be the family device, which I am taking to be $Oracle(T_{family})$. Indeed, in consequence of the fundamental role played by the the type T_{bm} in the family device, we have the saliency relation:

$$Oracle(T_{family}) \mid\!\!\vdash T_{bm}(\dot{b}, \dot{m}) \Rightarrow T_{is}$$

In particular, familiarity with the family device involves the capacity to recognize situations as being of type T_{bm}.

Consider now the listener in Sacks, 'baby–mommy' scenario, who hears the two sentences

The baby cried. The mommy picked it up.

To a native speaker of English, this clearly describes a situation, s, (possibly a fictional one) of type T_{bm}, (in situation-theoretic terminology, s is the *described situation*). Thus, under normal circumstances, the speaker's choice of words and the listener's interpretation of those words are governed by the constraints salient in the family device, in particular the constraint

$$T_{bm}(\dot{b}, \dot{m}) \Rightarrow T_{is}$$

mentioned above. It is this constraint that, for the speaker, leads to the two actors being preferentially described as 'baby' and 'mommy' and, for the listener, leads to the reference of 'the mommy' being heard as the mommy of the baby. And it is this constraint that has, as a derivative, the associative link

$$T_{baby} \Rightarrow T_{mother}$$

As the above discussion indicates, it is possible to dispense with associative links altogether, and work entirely with situation-type constraints such as $T_{bm} \Rightarrow T_{is}$. However, the result of an evolved attunement to such constraints is that members do develop such connections *between objects*, which is to say that they have informational status, which a theory of information ought to reflect. In infon form, the above associative link is written

$$\ll \text{linked}, T_{baby}, T_{mother}, t_{now}, 1 \gg$$

Notice also that normal use of (normative) constraints of the kind under discussion is often unconscious and automatic. Reliance on such constraints is a matter of attunement rather than awareness.

Familiarity with the type T_{bm} also involves attunement to the *category binding* constraint

$$\text{crying} \Rightarrow T_{baby}$$

that links the activity of crying to the category of babies. Again, it would be possible to avoid the introduction of such constraints, and to work instead with situation-type constraints (in the case of this example, going through the type T_{bm}), but again they reflect more closely the basic manner in which these connections figure in the everyday cognitive activity of members, and so it seems sensible to include them. As an informational item, the binding of crying to the category of babies would be written

$$\ll \text{category-bound-to}, \text{crying}, T_{baby}, t_{now}, 1 \gg$$

Notice that the account developed above concerns the relevant information structures 'in the world'. That is to say, I have been trying to identify the informational links that a member will (unconsciously) make use of in order to understand a certain situation (either observed or described). I have not attempted to describe the member's cognitive processes in any way. This could be done. For instance, as part of the process of evolving the type T_{bm}, a member will learn that one of the two individuals concerned is usually referred to as a 'baby' and the other as a 'mother'.

One final remark to conclude this section: the account of normality just developed is very much a contextual one. What I describe is normal behavior *in a given context*. Moreover, that context may be an *evolved* one, built up by a process of the member repeatedly interacting with the environment.

7 So Whose Theory Is It?

Writing in (Psathas 1979), Heritage and Watson (pp. 123–124) set out the central task of the ethnomethodologist as follows:

A central focus of ethnomethodological work is the analysis of the practical sociological reasoning through which social activity is rendered accountable and orderly. Assumed by this concern is the notion that all scenic features of social interaction are occasioned and established as a concerted practical accomplishment, in and through which the parties display for one another their competence in the practical management of social order. As analysts, our interest is to explicate, in respect of naturally occurring occasions of use, the methods by which such orderliness can be displayed, managed, and recognized by members.

The question at once arises, how does one, as Heritage and Watson put it, "explicate ... the methods by which such orderliness can be displayed, managed, and recognized by members"? How does one describe the 'social structure' that Garfinkel, Sacks, and others claim governs the way members encounter the world and interact with one another?

A commonly heard view is that such an analysis of everyday (sociolinguistic) phenomena can only be carried out using the linguistic tools of natural language, relying upon the linguistic competence and everyday experience of both producer and reader of such an analysis in order to capture and describe various features of the observed data. To repeat part of the quotation from Benson and Hughes given in Section 1: "... an important constraint on sociological inquiry is that the categories, the concepts used, and the methods for using them, must be isomorphic to the ways they are used in common-sense reasoning."

But an examination of, say, the Sacks' article considered in this paper will indicate that neither the language nor its use is in any sense 'everyday'. Rather, Sacks adopts (indeed, in his case, helped to develop) a specialized and highly stylized linguistic *genre*, whose complexity varies inversely with the 'everydayness' of the particular phenomenon under discussion. Whether or not there is, nevertheless, an 'isomorphism' between such ethnomethodological use of language and everyday reasoning is of little consequence, since there is quite evidently no *natural* isomorphism. The production and proper understanding of an analysis such as Sacks' involves considerable effort, including the mastery of an entire linguistic *genre*, and an associated ontology, quite unlike 'everyday English' or 'everyday concepts'.

The reason for this linguistic complexity lies in the fact that the goal of the analysis is the identification and description of formal, abstract structures, and everyday language is not well-suited to such a task. Mathematics, on the other hand, is very well-suited to this kind of purpose, since mathematics is, by definition, the study of formal, abstract structures. As a piece of mathematics, the situation-theoretic analysis of the Sacks data presented here (or in (Devlin and Rosenberg, to appear)) is extremely simple—indeed, a mathematician would describe it as essentially 'trivial' (a term that mathematicians generally use in a technical, non-perjoratory way).

But this triviality and appropriateness of the technical, descriptive machinery provided by situation theory comes at a price: it is only easily and readily accessible to those sufficiently familiar with (enough) mathematics. Though by no stretch of the imagination 'everyday use of language', the descriptive machinery of ethnomethodologists such as Sacks is *grounded* in everyday language, and thus is, in principle, accessible to any sufficiently competent English speaker, given only an adequate motivation and enough time. (It took me, a mathematician hitherto unfamiliar with sociology, some six months of fairly intensive effort, including the assistance of Rosenberg, a social scientist familiar with the ethnomethodological literature, before I felt I was able to understand Sacks' analysis reasonably well.) Given the nature of elementary and high school education prevalent through most of the western world, where what 'mathematics' is taught is largely quantitative and algorithmic (as opposed to qualitative and descriptive), for most social scientists, the time, and effort, required to achieve an adequate mastery of the appropriate descriptive tools from mathematics is considerably longer, and indeed may be, for many, inaccessibly long.

However, this is an empirical consequence of the prevailing educational system, not an intrinsic feature of descriptive mathematics. To anyone sufficiently well-versed in mathematics, and familiar with mathematical ontologies, a description carried out with the aid of mathematics is just that: a description. Indeed, to such a person, the situation-theoretic analysis of the Sacks data presented above may well be (and for myself certainly was) much simpler to produce and to understand than a more traditional, ethnomethodological analysis such as Sacks' own. To someone well-versed in mathematics, the use of descriptive and analytic mathematical tools, as in this paper, is a perfectly *natural*, and indeed appropriate, medium for carrying out an investigation of the abstract structures involved in any phenomenon. Such an analysis will be open to inspection and validation by others with a similar background.

There is then, I maintain, no intrinsic difference between the kind of analysis developed by Sacks and the situation-theoretic analysis of the same data developed by Rosenberg and myself. The difference lies in the groups to which the two analyses are, first, accessible and, second, natural. It would probably be too naive to expect that social scientists would rush to learn, and embrace, situation-theoretic, or other descriptive-mathematical, techniques, though I suspect that in due course some will. Much more likely is that increasing numbers of computer scientists, already familiar with the language of mathematics, will find themselves becoming involved in ethnomethodological questions in their work on problems of systems design. If this does indeed turn out to be the case, then it is this group for whom the kind of analysis presented here will be most appropriate. Whether such studies are classified as 'ethnomethodological', or whether they will take their place alongside ethnomethodology as an alternative means of analysis, is of little importance. The crucial issue, surely, is to achieve greater understanding and wider utility of that understanding.

8 Situation Theory as Endogenous Logic

Expression of the need to develop descriptive-mathematical tools for the analysis of social phenomena is not solely restricted to mathematicians, such as myself, who are trying to understand such phenomena. Such a program has been set out by some working within the ethnomethodological tradition. In particular, in (Button 1991: 49), Jeff Coulter sets out the following *desiderata* for a 'logic of language' that is, as he put it 'adequately wedded to *praxis*':

(A) An extension of analytic focus from the proposition or statement, sentence or speech-act, to 'utterance design' or 'turn-at-talk'.

(B) An extension of analytic focus to encompass indexical expressions as components of *sequences* in terms of their logical properties and relations, especially their *inferential affordances*.

(C) A respecification of the concept of 'illocutionary act' to exclude *a prioristic* efforts to isolate 'propositional contents', and more fully to appreciate the socially situated availability of 'what an utterance could be accomplishing' *in situ*, especially in respect of its properties of design, sequential implication and turn-allocation relevances; in other words, its *interactionally* significant properties.

(D) A development of the concept of a *combinatorial* logic for illocutionary activities *in situ*.

(E) A development of an informal or endogenous logic for the *praxis* of person, place, activity, mental predicate and collectivity categorizations, and their interrelationships, among other domains of referential, classificational and descriptive operations. This requires *abandonment* of formal semantic theoretic schemes deriving from set theory, extensionalism, generative- (transformational) grammar, truth-conditional semantics, and componential analysis as resources.

(F) Abandonment of the preoccupation with 'correctness' defined as usage in accord with any rule specified *independently* of an analysis or orientation to ascertainable members' situated relevances, purposes and practices.

(G) Abandonment of *a priori* invocations of mathematical concepts in the analysis of the informal logic of reasoning and communication; only those concepts warranted by studies of actual, *in situ* practical orientations of persons may be employed.

(H) Replacement of the goal of logical regimentation in favor of logical *explication*.

(I) Awareness of the varieties and modalities of what could count as 'rules of use' of linguistic/conceptual resources.

(J) Abandonment of intellectual prejudices and generic characterizations concerning the putative 'vagueness', 'disorderliness', 'ambiguities', 'indeterminacies', 'imprecisions' and 'redundancies' or ordinary language use.

(K) Formalization, but not axiomatization, becomes an objective, but not necessarily the production of an *integrated system* of formalizations.

(L) Adherence to the constraint that formulations of rules of practical reasoning and communication be sensitive to *actual*, and not exclusively hypothetical, cases of *praxis*.

(M) Extension of the concept of a 'logical grammar' to encompass the diversity of phenomena studied as components of conceptual *praxis*, requiring the de-privileging of 'strict categoricity rules' and the fuller exploration of the ties between Logic and Rhetoric.

(N) De-privileging of all decontextualized standards for the ascription of 'rationality' and 'truth' without sacrificing their position as components of real-worldly reasoning in the arts and sciences of everyday affairs.

(O) Recognition of the priority of *pre*-theoretical conceptualizations of phenomena as constraints upon 'technical' renditions of them.

How close does situation theory come to fulfilling this 'sociologist's wish list' for a 'logic wedded to *praxis*'? The answer is 'very close indeed'. Indeed, not only are all the aims and guiding principles of situation theory consistent with Coulter's requirements, in many cases they are identical. In this section, I examine this issue, taking Coulter's requirements one at a time.

(A) An extension of analytic focus from the proposition or statement, sentence or speech-act, to 'utterance design' or 'turn-at-talk'.

Though situation semantics has hitherto concentrated almost exclusively on propositions, statements, sentences, and speech-acts, as a semantic theory, situation semantics focuses on particular *utterances* of expressions. So in part, Coulter's requirements have already been met by situation semantics. In so far as this is not the case, there is nothing to prevent situation theory being applied to 'utterance design' or 'turn-at-talk'. Indeed, both my paper with Rosenberg (Devlin and Rosenberg, to appear) and this present paper are further steps in this direction, and our forthcoming paper (Devlin and Rosenberg, in preparation a) and monograph (Devlin and Rosenberg, in preparation b) will carry this development further.

(B) An extension of analytic focus to encompass indexical expressions as components of *sequences* in terms of their logical properties and relations, especially their *inferential affordances*.

From its inception, situation semantics took the issue of indexicality very seriously, and requirement (B) could be said to be one of the motivating factors for situation semantics.

(C) A respecification of the concept of 'illocutionary act' to exclude *a prioristic* efforts to isolate 'propositional contents', and more fully to appreciate the socially situated availability of 'what an utterance could be accomplishing' *in situ*, especially in respect of its properties of design, sequential implication and turn-allocation relevances; in other words, its *interactionally* significant properties.

Though the notion of the 'propositional content' of a declarative utterance played a major role in the early work on situation semantics, subsequent developments led to propositional content becoming just one of a number of features of (certain kinds of) utterances. Indeed, in the brief discussion of the five speech-act categories of Searle presented in (Devlin 1991), the *impact* of utterances played a significant, and for some kinds of utterance the most significant, role. Indeed, situation theory appears to be well suited to fulfill the positive requirements listed under (C), and my work with Rosenberg in (Devlin and Rosenberg, to appear) and subsequently has been, and is, moving in this direction.

(D) A development of the concept of a *combinatorial* logic for illocutionary activities *in situ*.

Insofar as situation theory is a 'combinatorial logic', I suppose this aim is met, but it is not at all clear to me just what Coulter has in mind here.

(E) A development of an informal or endogenous logic for the *praxis* of person, place, activity, mental predicate and collectivity categorizations, and their interrelationships, among other domains of referential, classificational and descriptive operations. This requires *abandonment* of formal semantic theoretic schemes deriving from set theory, extensionalism, generative- (transformational) grammar, truth-conditional semantics, and componential analysis as resources.

This expresses very well what current situation theory is trying to achieve, and the means by which it is setting about that task. The inadequacy, and indeed the unsuitability, of set theory and classical logic as a foundation for situation theory was realized quite early on in the development of the subject, and has long since been abandoned in favor of a 'top down' approach that starts off with empirical observations of communication and cognition.

(F) Abandonment of the preoccupation with 'correctness' defined as usage in accord with any rule specified *independently* of an analysis or orientation to ascertainable members' situated relevances, purposes and practices.

As indicated above, as a domain-led development, situation theory has never had such a preoccupation.

(G) Abandonment of *a priori* invocations of mathematical concepts in the analysis of the informal logic of reasoning and communication; only those concepts warranted by studies of actual, *in situ* practical orientations of persons may be employed.

This echoes remarks I have made both in this paper and elsewhere that the task at hand requires new mathematical concepts, abstracted from the target domain.

(H) Replacement of the goal of logical regimentation in favor of logical *explication*.

Again, this is a goal I, and others, have expressed for situation theory.

(I) Awareness of the varieties and modalities of what could count as 'rules of use' of linguistic/conceptual resources.

This principle seems consistent with the development of situation theory to date.

(J) Abandonment of intellectual prejudices and generic characterizations concerning the putative 'vagueness', 'disorderliness', 'ambiguities', 'indeterminacies', 'imprecisions' and 'redundancies' of ordinary language use.

Situations are explicitly designed to capture, or help capture, each of these aspects of ordinary language use. See the discussion of situations in (Devlin 1991).

(K) Formalization, but not axiomatization, becomes an objective, but not necessarily the production of an *integrated system* of formalizations.

This is the substance of the stress Rosenberg and I placed upon the 'toolbox' approach to the development of situation theory for use in sociology.

(L) Adherence to the constraint that formulations of rules of practical reasoning and communication be sensitive to *actual*, and not exclusively hypothetical, cases of *praxis*.

Situation theory has always adhered to this constraint. Likewise, Coulter's final three requirements all seem to be among the principles that guide the development of situation theory, and I will not comment on them individually.

Though initially developed for a different purpose, or at least, for a purpose that many would have supposed was different, it seems then that situation theory comes extremely close to being the kind of theory Coulter has in mind. If 'logic' is taken to mean 'the science of reasoning and inference', as I argue for in (Devlin 1991), then situation theory could be said to be an attempt to develop an 'endogenous logic' in the sense of Coulter.

Though Coulter does not argue against a *mathematical* development of such a 'logic', many sociologists seem to think that a mathematical approach is incompatible with ethnomethodology. As I have argued elsewhere in this essay, for the most part such skepticism seems to be based on an erroneously narrow view of what constitutes mathematics. Admittedly, it may well not be possible to develop endogenous logic in the formal, mathematical fashion of, say, predicate logic. Indeed, it is, I think, unlikely that a single, uniform, mathematical framework will capture all of the features of everyday conversation and human action. But the development and use of mathematical tools *as part of* an 'endogenous logic' should surely provide the sociologist, on occasion, with an additional level of precision not otherwise available. Indeed, the fact is that the additional level of precision that mathematical techniques can afford enabled Rosenberg and myself to identify, and rectify, a number of points of unclarity in what is acknowledged to be an exemplary ethnomethodological analysis. This does not mean that ethnomethodology is likely to become a 'mathematical science' in the sense of physics and chemistry. But it does indicate, quite clearly, that mathematical techniques can make a useful addition to the arsenal of descriptive and analytic tools at the ethnomethodologist's disposal.

References

Barwise, J., and Perry, J.: *Situations and Attitudes.* Bradford Books, MIT Press, 1983.

Button, G. (ed.): *Ethnomethodology and the Human Sciences.* Cambridge University Press, 1991.

Devlin, K.: *Logic and Information*, Cambridge University Press, 1991.

Devlin, K., and D. Rosenberg: "Situation Theory and Cooperative Action," in: *Situation Theory and its Applications*, Volume 3, CSLI Lecture Notes, to appear.

Devlin, K., and D. Rosenberg: "Networked Information Flow Via Stylized Documents," (in preparation a).

Devlin, K., and D. Rosenberg: *The Logical Structure of Social Interaction.* (in preparation b).

Dretske, F.: *Knowledge and the Flow of Information.* Bradford Books, MIT Press, 1981.

Garfinkel, H.: *Studies in Ethnomethodology.* Prentice-Hall, 1967.

Phillips, D.: *Knowledge from What?* Rand McNally, 1971.

Psathas, G.: *Everyday Language: Studies in Ethnomethodology.* New York: Irvington, 1979.

Sacks, H.: "On the Analyzability of Stories by Children," in: J. Gumpertz and D. Hymes (eds.), *Directions in Sociolinguistics. The Ethnography of Communication.* Holt, Rinehart and Winston Inc. (1972) 325-345.

Printing: Weihert-Druck GmbH, Darmstadt
Binding: Theo Gansert Buchbinderei GmbH, Weinheim

Lecture Notes in Artificial Intelligence (LNAI)

Lecture Notes in Computer Science